Articular Cartilage Tissue Engineering

Synthesis Lectures on Tissue Engineering

Editor
Kyriacos A. Athanasiou, *University of California, Davis*

Articular Cartilage Tissue Engineering
Kyriacos A. Athanasiou, Eric M. Darling, and Jerry C. Hu
2009

Tissue Engineering of Temporomandibular Joint Cartilage
Kyriacos A. Athanasiou, Alejandro A. Almarza, Michael S. Detamore, and Kerem N. Kalpakci
2009

Engineering the Knee Meniscus
Kyriacos A. Athanasiou and Johannah Sanchez-Adams
2009

Articular Cartilage Tissue Engineering

Kyriacos A. Athanasiou, Eric M. Darling, and Jerry C. Hu

ISBN: 978-3-031-01450-5   paperback
ISBN: 978-3-031-02578-5   ebook

DOI: 10.1007/978-3-031-02578-5

A Publication in the Springer series
SYNTHESIS LECTURES ON TISSUE ENGINEERING

Lecture #3
Series Editor: Kyriacos A. Athanasiou, *University of California, Davis*

Series ISSN
Synthesis Lectures on Tissue Engineering
Print 1944-0316   Electronic 1944-0308

Articular Cartilage Tissue Engineering

Kyriacos A. Athanasiou
University of California, Davis

Eric M. Darling
Brown University

Jerry C. Hu
University of California, Davis

SYNTHESIS LECTURES ON TISSUE ENGINEERING #3

ABSTRACT

Cartilage injuries in children and adolescents are increasingly observed, with roughly 20% of knee injuries in adolescents requiring surgery. In the US alone, costs of osteoarthritis are in excess of $65 billion per year (both medical costs and lost wages). Comorbidities are common with OA and are also costly to manage. Articular cartilage's low friction and high capacity to bear load makes it critical in the movement of one bone against another, and its lack of a sustained natural healing response has necessitated a plethora of therapies. Tissue engineering is an emerging technology at the threshold of translation to clinical use. Replacement cartilage can be constructed in the laboratory to recapitulate the functional requirements of native tissues. This book outlines the biomechanical and biochemical characteristics of articular cartilage in both normal and pathological states, through development and aging. It also provides a historical perspective of past and current cartilage treatments and previous tissue engineering efforts. Methods and standards for evaluating the function of engineered tissues are discussed, and current cartilage products are presented with an analysis on the United States Food and Drug Administration regulatory pathways that products must follow to market. This book was written to serve as a reference for researchers seeking to learn about articular cartilage, for undergraduate and graduate level courses, and as a compendium of articular cartilage tissue engineering design criteria.

KEYWORDS

knee, hip, shoulder, articular cartilage, tissue engineering, chondrocyte, osteoarthritis, biomechanics, cartilage products, cartilage transplant, cartilage diseases/surgery, chondrocyte transplantation, autologous/adverse effects, United States Food and Drug Administration, device approval, orthopedic equipment/classification, cartilage epidemiology/physiopathology, incidence, athletic injuries/epidemiology, stem cells, bioreactors, direct compression, hydrostatic pressure, shear, micropatterning, single cell mechanics

To Thasos and Aristos. Always excelling is unavoidable.

Αφιερωμένο στους Θάσο και Άριστο. Αίεν αριστεύειν είναι αναπόφευκτο. $-KA^2$

To my parents, John and Marilynn, who provided a childhood of unconditional love, support, and encouragement, and to my wife, Louise, who has provided an adulthood of that and more. —EMD

To my family, our past and current students, and my friends back in Houston; I think of everyone on my treks, and wish that you can see what I see. —JH

Contents

Preface

Articular cartilage injuries, which are well known for their inability to heal, oftentimes degenerate inexorably to disastrous impairment. Multitudes of treatments have been devised for this age-long and vexing problem, but no satisfactory long-term solutions have been established. Over the past two decades, however, the swift growth and development of new knowledge and technologies for cartilage formation, pathology, and repair have been exciting, humbling, and inspirational. Tissue engineering, a young and vigorous field, is at the cusp of applying our understanding of biological systems and engineering platforms to clinical problems. These dynamic times have stirred in us the desire to pause and to survey the wealth of progress related to articular cartilage regeneration. This book highlights important historical and contemporary advances in the field of articular cartilage tissue engineering. Physiology, pathology, and current treatment options are presented, along with business and regulatory aspects of the development of cartilage products. It is our hope that undergraduate students, graduate students, and academic and industrial researchers alike will find this information useful in this time of rapid flux.

Kyriacos A. Athanasiou, Eric M. Darling, and Jerry C. Hu
University of California, Davis
September 2009

CHAPTER 1

Hyaline Articular Cartilage

hyaline (–adj): transparent or translucent; from Greek hyalos (ὕαλος), meaning glass

Glass-like in appearance, hyaline articular cartilage lines the ends of articulating bones. A tissue with low friction and high capacity to bear load, cartilage serves the critical function of permitting movement of one bone against another. Breakdown of this tissue results in significant pain, reduction and loss of mobility, and billions of dollars in medical costs and lost wages. Functional tissue engineering of cartilage endeavors to produce a solution that does not necessarily replicate the exact biological structure but instead creates a replacement tissue that functions as well as the healthy original.

Functional tissue engineering endeavors to produce a solution that does not necessarily replicate the exact biological structure but instead creates a replacement tissue that functions as well as the healthy original. To accomplish this goal, the composition, structure, and function of healthy tissues must first be well understood since the mechanical and biological attributes are tied directly to these characteristics.

1.1 COMPOSITION, STRUCTURE AND FUNCTION OF HYALINE CARTILAGE

Hyaline articular cartilage is composed of specialized proteins and macromolecules that allow the tissue to function in the rigorous mechanical environments of articulating joints, such as the knee, hip, and shoulder (Figure 1.1). Collagens and proteoglycans interact with a charged fluid environment to give articular cartilage its unique mechanical properties. This section discusses the biochemical composition of hyaline cartilage, as well as how it is organized at the micro- and macro-structural levels. The composition and structure of the tissue have a direct role in its function as a mechanical surface through regulation of its tensile, shear, and compressive properties.

1.1.1 BIOCHEMICAL COMPOSITION

Though articular cartilage is a metabolically active tissue that maintains its extracellular matrix in a state of constant turnover [1], not all molecular components are reconstituted at the same rate, and variations exist based on the spatial location within the tissue. Degradation and synthesis are concentrated in the regions immediately surrounding chondrocytes rather than in the territorial and interterritorial regions of the tissue [2]. The turnover of collagen is estimated to be very slow (>100 years), whereas aggrecan turnover is more rapid, with a half-life of 8-300 days in rabbits [3].

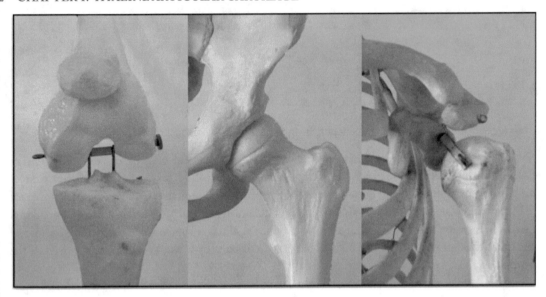

Figure 1.1: The knee, hip, and shoulder joints shown in a model skeleton.

The composition of articular cartilage changes as the tissue develops. However, mature articular cartilage is composed primarily of water, approximately 70-80% by weight. The solid fraction of the tissue is primarily collagens (50-75%) and proteoglycans (15-30%) (Figure 1.2), with the remaining balance including minor protein molecules and chondrocytes [1, 4]. This mix of collagens and proteoglycans form an integrated network that provides the basis for the mechanical properties observed in articular cartilage.

1.1.1.1 Synovial Fluid

Water is the main fluid component in articular cartilage, as well as the synovial fluid that is present in the joint capsule. Inorganic salts such as sodium, potassium, calcium, and chloride are also present in synovial fluid in appreciable amounts. Most of the water in cartilage tissue is contained in the molecular pore space of the extracellular matrix but also permeates throughout the entire tissue. Since cartilage has no vascularity, the chondrocytes obtain nutrients through diffusion from the joint space. As the primary carrier, interstitial fluid plays an important role in transporting both nutrients and waste within the tissue [5, 6].

Fluid permeating the cartilage matrix also has an important mechanical role. Compressive loading is a constant stressor of articular cartilage, and without a high water fraction, the tissue would break down much more quickly under constant use. Compressive loads can force the fluid from the tissue, and over the course of a day, effectively decrease the total water fraction. However, for short periods of loading and unloading the frictional resistance between the water and solid matrix requires high pressure to cause interstitial fluid flow [1]. Functionally, mechanical compression of cartilage

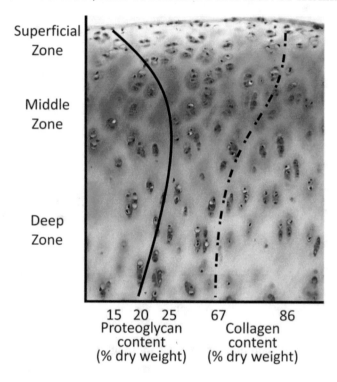

Figure 1.2: Proteoglycans and collagen contents vary throughout the zones of cartilage.

causes rapid pressurization of the fluid in the tissue, which in turn supports the load (Figure 1.3). This mechanism allows for the longevity of cartilage under repeated compression since loading is borne by fluid instead of a solid-solid interaction [7, 8].

1.1.1.2 Collagens

Collagens serve a primary role in the structure of connective tissues throughout the body. They are comprised of repeating amino acid sequences (glycine, proline, hydroxyproline, etc.) and exhibit a characteristic triple helix structure. Collagen type II is the predominant collagen type in articular cartilage, comprising over half the dry weight of the tissue [9]. Collagen fiber orientation varies through the depth of articular cartilage with the superficial zone containing tangentially arranged fibers, the deep zone containing radially oriented fibers, and the middle zone having both an arcade-like structure and randomly oriented fibers that forms the transition between the other zones [10] (Figure 1.4).

Hyaline cartilage also contains other fibrillar and globular collagen types, such as types V, VI, IX, and XI [11]. While the definitive roles of these other collagen types are not fully known, they are believed to play a role in intermolecular interactions as well as modulating the structure of collagen

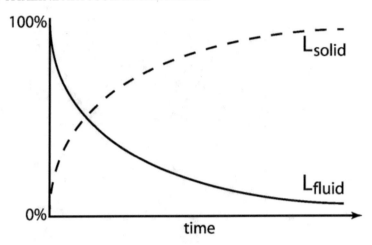

Figure 1.3: Initially, load applied onto cartilage is borne almost exclusively by the fluid phase (L_{fluid}). As the fluid exudes from cartilage, the solid matrix begins to bear more of the load (represented by L_{solid}). As time goes on, load borne by the fluid phase approaches zero.

type II [1]. Responte and Athanasiou give a more thorough treatment of the role of collagens in articular cartilage [12]. For example, collagen type IV is found primarily in the pericellular matrix and may contribute to the mechanical function of the chondron (the combined cell-pericellular matrix structure) and/or regulate interactions between the chondrocyte and extracellular matrix [13, 14]. Collagen type X, found primarily in the zone of calcified cartilage, appears to play a role in cartilage mineralization at the interface between cartilage and the underlying bone [15].

1.1.1.3 Proteoglycans

Proteoglycans are large macromolecules comprised of a protein core with attached polysaccharide chains (glycosaminoglycans). The primary proteoglycan in articular cartilage is aggrecan, which consists of a hyaluronana core with numerous glycosaminoglycan side chains (Figure 1.5). The dominant polysaccharides in this macromolecule are chondroitin and keratan sulfates in mature articular cartilage. The protein core contains several distinct globular and extended domains where glycosaminoglycans attach. Hyaluronan binds non-covalently through one such domain and is stabilized by a link protein [16]. The conglomeration of many proteoglycans into large macromolecules is critical for proper functionality of cartilage tissue.

Proteoglycan networks in articular cartilage can be thought of as a mesh that is interlaced throughout, within the more organized collagen structure. Aggrecan molecules are bound to a single, long chain of hyaluronan to form large proteoglycan aggregates, which result in an overall molecular weight of $50\text{-}100 \times 10^6$ Daltons [17, 18]. The large size of this polymer mesh acts to immobilize and restrain it within the collagen network. The presence of carboxyl and sulfate groups gives proteoglycans a negative charge, which in turn gives cartilage extracellular matrix a

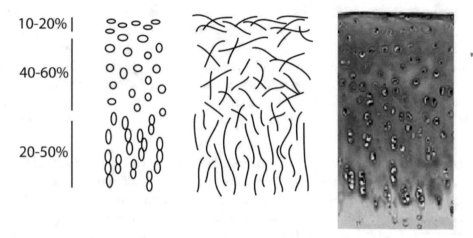

10-20%

40-60%

20-50%

Figure 1.4: Both cells (right) and collagen fibers (middle) are organized within cartilage into superficial, middle, and deep zones, consisting of 10-20%, 40-60%, and 20-50% of the overall tissue depth, respectively.

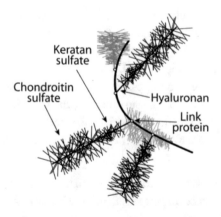

Keratan sulfate

Chondroitin sulfate

Hyaluronan

Link protein

Figure 1.5: The negatively charged keratan and chondroitin sulfate on aggrecan macromolecules repel each other and flare like tube brushes.

net negative charge known as a "fixed charge density" [19]. Because of this charge, the matrix imbibes fluid, swelling the tissue to maintain equilibrium. The swelling is balanced against the elastic restraint of the collagen network [20]. As mentioned in the section on synovial fluid, the functional properties of cartilage under compression are highly dependent on fluid pressurization within the tissue. Since the presence of proteoglycans assists in the imbibition of water, it is apparent that a loss of proteoglycans can result in a lack of fluid pressurization, and therefore, improper mechanical function. This breakdown in functionality is seen in advanced stages of diseases like osteoarthritis.

Other proteoglycans within articular cartilage include biglycan, decorin, and fibromodulin, which also are comprised of core proteins with various glycosaminoglycan species attached as side chains [17]. As with the minor collagen types present in cartilage, the precise roles of these proteoglycans are not fully known. However, it is likely they assist in matrix assembly by associating with the collagen structure during development and repair.

1.1.1.4 Other Molecules

In addition to the proteoglycans and collagens, articular cartilage also contains a small fraction of non-collagenous proteins. These include fibronectin, cartilage oligomeric protein, thrombospondin, tenascin, matrix-GLA (glycine-leucine-alanine) protein, chondrocalcin, and superficial zone protein [1]. The functions of these molecules are currently investigated, toward a better understanding of the intricacies associated with cartilage performance. For example, superficial zone protein has been shown to play an important role in the surface properties of articular cartilage through either a lubricating or protective mechanism [21]. Other matrix constituents in articular cartilage include lipids, phospholipids, glycoproteins, and inorganic crystal compounds [22–25].

1.1.2   STRUCTURE

Hyaline cartilage has two primary reference frames when describing matrix organization. The first is with respect to depth within the tissue. Cartilage has a zonal structure that varies from the surface of the tissue through to the bone [26]. Variations exist in cell morphology, collagen fiber orientation, and biochemical composition (Figure 1.4). The second type of organization is observed at the microscale, with matrix structure and composition varying with respect to distance from the chondrocyte membrane. The tissue immediately surrounding the chondrocyte is termed the pericellular matrix, which is in turn surrounded by the territorial and interterritorial matrices (Figure 1.6). These outer regions are also termed, more generally, the extracellular matrix.

1.1.2.1 Zonal Structure

Successive zones exist in hyaline cartilage from the articulating surface down to the subchondral bone (Figure 1.4). These regions (superficial/tangential, middle/transitional, deep/radial, calcified) can be identified by extracellular matrix structure and composition, as well as cell shape and arrangement within the tissue. Variations also exist in their mechanical characteristics, which tie directly into the overall functionality of cartilage.

The surface of articular cartilage is covered by a very thin, proteinaceous layer termed the lamina splendens [28]. This acellular, primarily non-fibrous region has a thickness ranging from hundreds of nanometers to a few microns. The precise role of the lamina splendens is not known. Hypotheses have suggested: (1) the layer exists to facilitate low friction and to protect wear of the cartilage surface, (2) it forms due to gradual accumulation of proteins and molecules from the synovial fluid, or (3) it is a visual artifact that results from processing or imaging. Several recent studies using different techniques (confocal microscopy, SEM, AFM) have confirmed the existence of the lamina

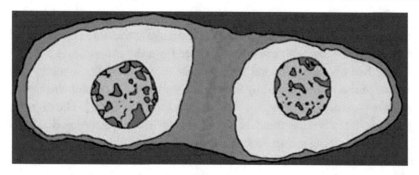

Figure 1.6: Chondrocytes (light gray) and their surrounding pericellular matrix (medium gray) form a functional unit termed the chondron. The adjacent territorial and interterritorial matrices (dark gray) are collectively termed the extracellular matrix [26].

splendens [29–31], and researchers are now trying to understand how it forms and what its function might be.

The superficial, or tangential, zone of articular cartilage comprises the upper 10-20% of the tissue. It is characterized by having small diameter, densely packed collagen fibers that are oriented parallel to the cartilage surface [32]. The matrix has a relatively low proteoglycan content as well as low permeability [33, 34]. The cells in this layer are densely packed and exhibit flattened, discoidal shapes that are oriented along the neighboring collagen fibers in a tangential direction [35, 36]. Superficial zone cells secrete specialized proteins that are hypothesized to facilitate the wear and frictional properties of the tissue [21].

The middle, or transitional, zone occupies approximately 40-60% of the total tissue thickness. The collagen fibers in this region exhibit an arcade-like structure interspersed with randomly oriented fibers [26]. Proteoglycan content reaches its maximum in the middle zone [37] (Figure 1.2). The cell density is much lower in this region than the superficial zone, and the cells themselves are more spherical in shape [35, 36].

The deep zone is the last region of purely-hyaline tissue before reaching bone. Its collagen structure is characterized by large fibers that form bundles oriented perpendicular to the articular surface and are anchored in the underlying subchondral bone [33]. Proteoglycan content is much lower than in the middle zone [37], and the cell density is also the lowest of the three cartilaginous zones [35]. Cells in the deep zone often group together in a columnar organization. They are slightly elongated and oriented in the direction of collagen fibers, perpendicular to the articular surface [36].

A thin line termed the "tidemark" is present between the deep zone and calcified zone of articular cartilage [38]. The calcified zone is a region of the tissue that transitions into the subchondral bone, minimizing the stiffness gradient between the rigid bone and more pliable cartilage [39]. Underlying this region of the cartilage is the subchondral bone, which is the ultimate anchorage point for cartilage tissue as a whole.

1.1.2.2   Territorial Structure

In addition to the zonal organization associated with cartilage extracellular matrix, the tissue also has a microscale structure oriented with respect to distance from the chondrocyte cell membrane (Figure 1.6). As described above, the region immediately surrounding a chondrocyte is termed the pericellular matrix and is characterized by having fine collagen fibers, high concentrations of proteoglycans, and the presence of fibronectin and collagen type VI [40, 41]. The exact function of the pericellular matrix is not fully understood. However, strong evidence indicates that it helps to protect the physical integrity of articular chondrocytes during compressive loading [27].

The region immediately surrounding the pericellular matrix (the territorial matrix) is composed of similar molecular constituents as the surrounding extracellular matrix, namely collagen type II and proteoglycans. In normal cartilage, collagen type VI has also been shown to be localized to this region [42]. The territorial matrix exhibits a higher concentration of proteoglycans than the surrounding extracellular matrix, as well as having a finer collagen structure [26].

The interterritorial matrix, which contains large collagen type IV fibers and varying concentrations of Proteoglycans, comprises the bulk of articular cartilage, providing the tissue with its mechanical properties. Loading of articular cartilage involves force transmission through the interterritorial, territorial, and pericellular matrices before reaching the chondrocytes. These regions likely assist in modulating strains seen at the cellular level [13]. The interterritorial regions are representative of bulk extracellular matrix tissue and contain large collagen type II fibers as well as varying concentrations of proteoglycans, dependent on depth from the surface. Therefore, structural breakdown of any region can dramatically affect the forces experienced by individual cells.

1.1.3   FUNCTION

The primary role of articular cartilage is to provide a low-friction, wear-resistant surface that can withstand large loads over decades of constant use. Within the body, cartilage functions to facilitate load support and load transfer while allowing for translation and rotation between bones. The degree of loading in an articulating joint is dependent on its location in the body. The force exerted on the hip has been calculated to be 3.3 times a person's bodyweight. The knee experiences a load of approximately 3.5 times bodyweight, the ankle 2.5 times bodyweight, and the shoulder 1.5 times bodyweight [43]. Experimentally, compressive stresses in the hip routinely reach 7-10 MPa and have been measured up to 18 MPa during more stressful activities such as standing up [44]. The biochemical and mechanical characteristics of articular cartilage directly affect how it performs in the joint. Changes in these characteristics can dramatically alter the loading profile, thereby beginning a process of degradation that can eventually result in total loss of the tissue

The deformation characteristics of articular cartilage play an important role in its mechanical functionality. The time- and rate-dependent behavior of articular cartilage stems from interstitial fluid flow through the solid matrix and is manifested via creep, stress relaxation, and energy dissipation or hysteresis [8, 45–47]. Sudden loading is initially borne by the fluid phase of the cartilage, helping to absorb the energy of impact that would otherwise be felt by the solid phase. The contact stress experienced by articular cartilage is also decreased as the tissue deforms upon loading since the area of contact between surfaces increases.

Articular cartilage is a highly complex material, essentially a fluid-saturated, fiber-reinforced, porous, permeable composite matrix [48]. The material properties of cartilage can be described as viscoelastic (time- or rate-dependent), anisotropic (dependent on orientation), and nonlinear (e.g., dependent on magnitude of strain) [1]. Mathematical models have been developed that attempt to describe these properties, which are dependent on the interaction among the different phases in cartilage (solid, fluid, ionic). Some of these models will be discussed in the following sections.

Loading and deformation of articular cartilage generate a combination of tensile, compressive, and shear stresses within the tissue [1]. Healthy cartilage can withstand many decades of rigorous use without deterioration or failure. This section describes the primary mechanical forces acting on articular cartilage.

1.1.3.1 Compression

Compressive loading is one of the primary types of mechanical stress experienced by articular cartilage. The compressive aggregate modulus, H_A, of cartilage ranges from 0.08 to 2 MPa and varies by depth in the tissue, location on the joint, and species [49–53]. Compression of cartilage is governed primarily by the movement of fluid through the interconnected pore structure of the solid matrix. Hydraulic permeability is a measure of fluid movement through a solid matrix and is related to pore size, structure, and connectivity [48,54,55]. The frictional drag associated with interstitial fluid flow through the porous, permeable solid matrix is the dominant dissipative mechanism for cartilage [47]. However, this functionality might be significantly impaired if the tissue structure is disrupted or otherwise not intact.

Cartilage has a low permeability and resists fluid flow, resulting in high drag forces as the interstitial fluid moves through the solid matrix. Interstitial fluid pressures are very large under compression, which is a significant mechanism for load support in the joint [48,54–57] (Figure 1.3). As fluid redistributes within the tissue, time-dependent changes occur in fluid movement which contribute to the viscoelastic properties of the tissue. As fluid pressure decreases over time, more load is supported by the solid fraction of the matrix, giving rise to creep and stress relaxation behaviors [58].

Volumetric changes occur as fluid is extruded from the tissue under compression. Upon removal of the compressive load, cartilage tissue recovers its initial dimensions. This is possible through the combination of the elasticity of the solid matrix and the imbibition of surrounding fluid [6]. Compression and recovery occur repeatedly at the microscale level during normal joint movement. However, over the course of a day, the bulk cartilage tissue is compressed slightly compared to its initial state. This results in total compressive strains of 15-20% [59]. However, a good period of inactivity (e.g., a night's sleep) will allow the tissue to fully recover.

1.1.3.2 Tension

Tension can occur when two cartilage surfaces slide across one another and pull in a single direction. At the surface of the tissue, tension also occurs when the cartilage is compressed, pulling the surrounding regions toward the point of loading. As cartilage is loaded in tension, the collagen fibrils within the solid matrix align and stretch along the axis of loading. The frictional properties of the tissue should help limit the magnitude of tensile strains, but small, repeated periods of tensile loading still occur during normal joint movement.

The tensile properties of cartilage are nonlinear due largely to the behavior of collagen fibers in the tissue. For small deformations, collagen fibers realign in the direction of loading. As tension increases, the crosslinked collagen fibers themselves begin to stretch, which results in the tissue exhibiting a higher stiffness at larger strains [60–62]. Cartilage exhibits linear equilibrium stress-strain behavior up to 15% strain [63].

The tensile Young's modulus is essentially a measure of the solid collagenous matrix and varies by both depth and orientation in the tissue [61]. Other contributing physical parameters include collagen fiber density, fiber diameter, amount of crosslinking, and the strength of ionic bonds and frictional interactions between the collagen and proteoglycan networks [64, 65].

The tensile modulus of healthy human cartilage varies from 5 to 25 MPa depending on the location on the joint and depth in the tissue [60, 63, 66]. In general, the superficial zone of cartilage is stiffer in tension than the middle or deep zones [61]. Furthermore, the upper regions of cartilage tissue are also stiffer when oriented along split-lines (the predominant collagen fiber orientation at the cartilage surface) [64, 66].

The viscoelastic behavior of cartilage is also dependent on interactions between the collagen and proteoglycan networks. Enzymatic extraction of glycosaminoglycans has been shown to effect a significant increase in collagen fiber alignment, which alters the rate of deformation, or creep, for cartilage samples under tension [65]. While the collagen-proteoglycan interactions appear to affect rate changes in the deformation of cartilage tissue, the intrinsic stiffness of the solid collagenous matrix is what primarily contributes to the stress-strain behavior and failure properties of cartilage in tension [1].

1.1.3.3   Shear

Articular cartilage undergoes shear through the depth of its tissue from normal rotational and translational movement in the joint. Physically, pure shear causes no compression of the tissue, only stretching, which is relatively independent of fluid pressurization [45]. Shear properties at equilibrium help to characterize the interaction among solid components in cartilage, without having to account for the contribution from fluid flow effects. The equilibrium shear modulus for articular cartilage has been found to vary from 0.05 to 0.25 MPa [67].

Shear testing in cartilage is often measured by applying an oscillatory torsional strain over a range of frequencies. This results in a dynamic shear modulus (G^*), which indicates the stiffness of the matrix, and a loss angle (δ), which indicates the dissipation of internal friction. A perfectly elastic material would have a loss angle of $0°$, whereas a perfectly viscous fluid would have an angle of $90°$. The shear behavior of cartilage is due to the interaction between collagen fibers and proteoglycan networks. Dynamic shear moduli for human, bovine, and canine samples vary from 0.1 to 4 MPa, with the loss angle being ∼10° [45,68–70]. These values are higher than the equilibrium modulus due to the transitory nature of the loading, which is normal for viscoelastic materials.

1.1.3.4 Friction

Friction is a measure of the resistive force that exists when two contacting surfaces move laterally relative to each other. Several different mechanisms have been proposed to explain the low friction values between cartilage surfaces; these include squeeze film lubrication [71,72], elasto-hydrodynamic lubrication [73,74], boundary lubrication [75–77], and fluid pressurization [78–81]. Current findings indicate the latter is the most influential mechanism. Experiments focusing on interstitial fluid pressurization show that as pressure decreases ten-fold, the coefficient of friction, which represents the ratio of the tangential frictional force to the compressive force, increases 250-fold. However, normal loading does not allow for interstitial pressures to drop this dramatically, so friction at the cartilage interface remains minimal. The coefficient of friction for a cartilage on cartilage interface (∼0.005) is lower than any other known bearing [82].

1.2   ARTICULAR CHONDROCYTES

When compared to other tissues in the body, articular cartilage is sparsely populated by cells. The chondrocyte, as the sole cell type resident within hyaline cartilage, is pivotal for the maintenance of the tissue. All chondrocytes within articular cartilage share common traits with respect to gene and protein expressions, surface markers, and cell metabolism. However, some differences do exist in the genetic, synthetic, and mechanical characteristics of cells with respect to their zone of origin in the tissue [83–86].

The chondrocyte is the basic metabolic unit of cartilage, and is responsible for limited matrix remodeling [3]. Since articular cartilage is avascular, chondrocytes obtain nutrients by diffusion from the synovial fluid, facilitated during joint movement [87]. Though chondrocytes have been categorized as all belonging to the same phenotype, transient metabolic differences among chondrocytes of different sizes [88] and zonal affiliations [83,87,89–91] have been observed *in vitro*. For instance, superficial zone chondrocytes were found to attach to tissue culture plastic slower than those from the deeper zones [90]. Deep zone cells displayed a higher label for vimentin [92], which has been hypothesized to resist compression of the cell [93,94]. Keratin sulfate synthesis has been observed to gradually increase through cartilage depth [83,87,89–91].

The characteristic gene and protein expressions of chondrocytes are closely associated with the matrix constituents of articular cartilage. Maintenance of the surrounding matrix requires synthesis of proteoglycans and collagens (described in a previous section) as well as other small molecules. Disease and injury can alter cartilage physiology as well as tissue turnover, which can progressively accelerate tissue breakdown. Compounding the problem is the sparse cell population's inability to repair the cartilage to any extent [95].

Articular chondrocytes from the superficial, middle, and deep zones have morphologies and expression profiles specific to their regions within the tissue. Cell diameters range from 10-13 μm, with superficial zone cells being smaller than middle/deep zone cells [83]. The morphology for cells near the surface of the tissue is flattened and discoidal, whereas in the middle zone the cells are more rounded and, in the deep zone, the cells are ellipsoidal and organized in columns perpendicular to the surface [42]. In general, middle/deep zone cells possess greater synthetic capabilities for the major molecular constituents of cartilage than superficial zone cells when cultured *in vitro* [83, 86, 87]. Chondrocytes from these two zonal populations also have different mechanical properties. The Young's modulus of superficial zone chondrocytes is approximately twice that of middle/deep zone chondrocytes (E_Y = 460 vs. 260 Pa, respectively) [85]. A similar relationship is seen for other elastic and viscoelastic properties as well. These variations are likely caused by the different strain levels that cells experience within the zones of cartilage. Tissue near the surface is compressed more than that in the bulk of the cartilage [13, 49], and hence, those cells might need to be stiffer to survive the high strains.

Chondrocytes are known to lose their phenotypic markers *in vitro*, as evidenced by a temporal loss of morphologic characteristics and changes in metabolic activities of cells when cultured in monolayers [90, 96]. However, chondrocytes cultured in agarose retain morphological and proteoglycan synthesis characteristics [83, 87, 89, 97]. This is likely due to the constrained three-dimensional environment, which forces a rounded morphology on the cells. Alternatively, chondrocytes cultured in monolayer flatten over the course of days and begin to proliferate, rapidly losing their characteristic expressions [96]. Three-dimensional culture in a hydrogel or similarly constraining material is hypothesized to facilitate the synthesis of cartilage-specific molecules. Additionally, the application of mechanical stimuli such as stress, strain, and pressurization can affect their phenotypic expressions through a phenomenon termed mechanotransduction [98] (Figure 1.7).

1.3   CHAPTER CONCEPTS

- Hyaline articular cartilage is a glass-like tissue that is avascular, aneural, and alymphatic.

- Articular cartilage is composed largely of water, collagen, proteoglycans, and cells. These components are arranged into zones and vary accordingly.

- Articular cartilage is composed of 70-80% water (per ww), 50-75% collagen (per dw), and 15-30% proteoglycans (per dw).

- Collagen type II is most abundant in hyaline articular cartilage, with collagen types V, VI, IX, and XI also being present.

- Charged proteoglycans cause the tissue to imbibe and retain water. Physiological compressive forces are borne and dissipated as the water is forced out of the tissue.

- The synovial fluid reduces friction and plays an important role in transporting both nutrients and waste within the tissue.

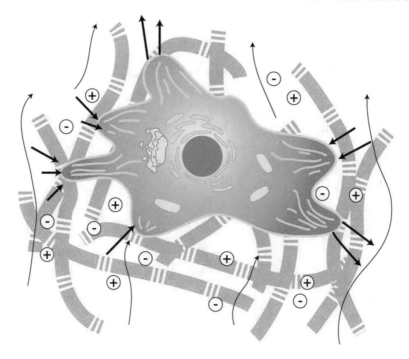

Figure 1.7: Mechanotransduction can occur via pulling, compressing, and shearing the cells (bold arrows). Other stimuli can include streaming potentials (fine arrows).

- Chondrocytes, the metabolic unit of cartilage, reside in lacunae, which has high collagen type VI contents.

- The lamina splendens, consisting of collagen fibers oriented along the direction of stress, covers the cartilage surface and serves to both resist shear and entrap the ECM within the tissue.

- The force exerted on the hip has been calculated to be 3.3 times a person's bodyweight. The knee experiences a load of approximately 3.5 times bodyweight, the ankle 2.5 times bodyweight, and the shoulder 1.5 times bodyweight.

- The compressive aggregate modulus, H_A, of cartilage ranges from 0.08 to 2 MPa and varies by depth in the tissue, location on the joint, and species. On average, the aggregate modulus is around 800 kPa.

- The tensile modulus of healthy human cartilage varies from 5 to 25 MPa. Thus, it should be remembered that the tensile modulus is about one order of magnitude higher than the compressive modulus.

CHAPTER 2

Cartilage Aging and Pathology: The Impetus for Tissue Engineering

This chapter examines how cartilage develops, and it then describes tissue changes that occur with age and pathology. As explained, cartilage inherently lacks an adequate healing response, motivating efforts in tissue engineering. Cartilage healing is difficult and deceptive. It is difficult because chondrocytes do not mount a sufficient healing response. It is also deceptive because, depending on the type of injuries, temporary functional restoration can last for years before the mechanically inferior repair tissue degenerates to result in significant pain and even disability. What is clear is that long-lasting functional restoration is naturally absent, and, if one judges "healing" by this criterion, then cartilage is devoid of a complete healing response.

2.1 CARTILAGE FORMATION

As animals with bilateral symmetry, human embryos develop three germ layers during embryogenesis. In between the ectoderm and endoderm is the mesoderm, from which cartilages arise. We will begin our discussion of cartilage formation from this point. Ossification will also be briefly discussed in relation to chondrogenesis, as knowledge on how cartilage calcifies may aid in developing methods to prevent unwanted calcification in tissue engineering efforts.

2.1.1 CHONDROCYTE CONDENSATION AND DIFFERENTIATION

From the mesoderm, the axial skeleton forms from the somites, the lateral plate mesoderm generates the limbs, and the craniofacial cartilages arise from the neural crest. Examples of axial skeleton cartilages include rib, intervertebral disc, and facet joint cartilages. The temporomandibular joint cartilages (i.e., disc, condyle, and fossa), along with auricular and nasal cartilages, are examples of craniofacial cartilages that are retained in the adult. One of the most essential articulations for maintaining quality of life is the temporomandibular joint. Diseases associated with this joint pose significant costs both financially and in terms of morbidity; a separate book in this series has been devoted to its treatment through tissue engineering [99]. The majority of the information provided below, though derived from many studies using cells of neural crest origin has been generalized to cartilage formation in the lateral plate mesoderm, i.e., cartilages of the knee, hip, and ankle.

In the sclerotome (the most medial segment of the somites after splitting into three segments) and the mesenchyme, cells first commit to becoming cartilage cells, causing the surrounding cells to express Pax1 and scleraxis. These two transcription factors then activate cartilage-specific genes [100, 101]. The cells condense into nodules and differentiate into chondrocytes. The chondrocytes divide rapidly and secrete cartilage-specific matrix, forming cartilage tissue (Figure 2.1).

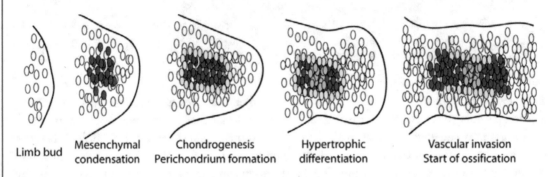

Limb bud   Mesenchymal   Chondrogenesis   Hypertrophic   Vascular invasion
            condensation   Perichondrium formation   differentiation   Start of ossification

Figure 2.1: Condensation and chondrogenesis. Cells first condensate (red) and differentiate into chondrocytes. Boundaries are defined as the perichondrium forms (in yellow), and chondrocytes can further undergo hypertrophy (green cells) to eventually mineralize to result in bone formation through endochondral ossification.

In response to sonic hedge hog protein (shh), bone morphogenetic proteins (BMPs) regulate Hox genes [102, 103] to initiate mesodermal cell proliferation and differentiation [104]. Msx-1 and Msx-2 are also involved at this stage as transcriptional repressors [105, 106]. The differentiation is additionally mediated by epithelial-mesenchymal interactions by TGF-β [107, 108]. TGF-β participates in condensation by regulating fibronectin [109], which in turn regulates N-CAM [110], which was initially thought to be required in maintaining condensation along with the adhesion molecule N-cadherin [111–113]. Note that N-CAM is not required in initiating condensation [110], but instead in its maintenance. Also, recent data obtained using organ culture have shown that N-cadherin-deficient limb buds were capable of mesenchymal condensation and chondrogenesis [114], indicating that N-CAM is not necessary in chondrogenesis, albeit an important player within normal cartilage development. Within the condensate, cell proliferation and adhesion is modulated by Hox genes (regulated by BMPs). Other transcription factors that regulate proliferation include CFKH-1, which regulates TGF-β, MFH-1, and osf-2, which is regulated by BMP-7 and vitamin D3. Syndecan, a receptor that binds to tenascin [115] and fibronectin [116], inactivates these to result in inactivated N-CAM, thus setting the boundaries for condensation. Around the condensate, noggin then binds to BMPs to slow or stop cell proliferation, halting condensate growth [117, 118]. From this point, the cells transition to differentiation via transcriptional activation of Hoxd-11-13 [119]. Within Figure 2.1, the molecular events that transpired above correspond to the point of chondrogenesis and perichondrium formation. This is a brief presentation of the intricate events that take place in condensation and differentiation. For a more in depth discussion, the reader should consult books and articles on development, such as Hall's excellent and thorough volume [120].

Many *in vitro* studies have manipulated the molecular players highlighted above to promote the accumulation of cartilage-specific matrix, both for developmental studies and for tissue engineering. As TGF-β1, β2, and β3 have been observed during differentiation *in vitro* [121], their efficacy has been examined in enhancing matrix production. The same is true for the BMPs. TGF-β has been shown as effective on cells that have not yet condensed while BMP-2 achieves similar effects after the cells have condensed or differentiated [121]. A more thorough treatment of select factors related to development and used in tissue engineering (e.g., TGF family, shh) will be presented in the next chapter, in Section 3.4.

2.1.2 HYPERTROPHY AND OSSIFICATION

Developmentally, cartilage can serve as a transition tissue to bone, and understanding this process may allow for its manipulation in tissue engineering. Osteogenesis can occur via intramembranous ossification, which is the direct conversion of mesenchymal tissue into bone, or via endochondral ossification, which is through the calcification of cartilage tissue [122]. Endochondral ossification occurs in both somatic and lateral plate cartilages to form, for example, the vertebrae and limbs, respectively, while stopping just short of facet joint and the articulating cartilages of the limbs. From the formation of cartilage tissue, the chondrocytes stop dividing and can undergo hypertrophy, during which the cells increase their volume [123]. The cartilage matrix is altered with the addition of collagen type X and increased fibronectin content. Collagen type X allows the tissue to become calcified, while VEGF, which transforms mesodermal mesenchyme cells to blood vessels, is secreted by the hypertrophic chondrocytes [124,125]. From here the blood vessels then infiltrate the cartilage, the hypertrophic chondrocytes die, and the cells that surrounded the cartilage become osteoblasts to make bone [126] (the last two steps of Figure 2.1). Our interest here lies just before this last step, on how the hypertrophy is regulated. That is, what are the factors that initiate it and, more pertinent to tissue engineered constructs, which factors prevent it from occurring.

It has been shown that hypertrophy follows chondrocytes switching from aerobic to anaerobic respiration. Evidence for this is provided by examining creatine kinase, an enzyme that catalyzes the formation of ATP in tissues under oxygen stress. Creatine kinase activity is related to both chondrocyte maturation and hypertrophy, as the activity of this enzyme increases to prepare for a hypoxic state [127]. Growth factors that have been shown to affect hypertrophy include the BMPs and TGF-β. BMP-2, -4, -6, and -7 have all been implicated in chondrocyte hypertrophy. Of these, BMP 6 and 7 are expressed in hypertrophic chondrocytes [128,129], and the exogenous addition of BMP-2 and -4 results in increases in chondrocyte hypertrophy [130,131]. Other factors include Rac1 and Cdc42, as overexpression of these small GTPases results in acceleration of hypertrophic differentiation [132]. Particularly interesting are the results Wu and associates [133] obtained with cyclic matrix loading. Collagen type X was shown to be up-regulated by stretch-induced matrix deformation, hinting that mechanical stimulation may also play a role in hypertrophy.

Exogenous methods to control hypertrophy and collagen type X deposition include the reduction of calcium concentrations [134]. TGF-β1/Smad3 signals inhibit hypertrophy [135], as well as FGF-2, which has been shown to reduce hypertrophy when added to the culture medium [136]. As Rac1 and Cdc42 overexpression has been shown to accelerate hypertrophy, the pharmacological inhibition of p38 signaling, which blocks the effects of Rac1 and Cdc42 overexpression, also reduces hypertrophy [132]. It has also been shown that blocking the α1β1-integrin prevents FXIIIA from inducing chondrocyte hypertrophy [137]. OP-1 is another candidate whose inhibition lowers the expression of collagen type X [138]. Parathyroid hormone-related protein has also been shown to have inhibitory effects on collagen type X expression [139]. Lastly, overexpression of Smad6 in chondrocytes results in delayed hypertrophy to the point of abolishing BMP-2's expected effect of hypertrophy induction [140]. Thyroxine's prevention of Meckel's cartilage from undergoing hypertrophy is also related to the topic at hand [141], though the cells in this case are derived from the neural crest. Whereas tissue engineered cartilages formed from differentiated chondrocytes have seldom been reported as undergoing calcification (except for select cases where chondrocytes from the calcified zone were used [142], the expansion of the field into stem cells would require a fine control and termination of cell differentiation. The methods listed above may be employed to prevent these cells from progress to hypertrophy.

2.2   AGING

Cartilages from skeletally immature, mature, and older patients display several prominent differences. While adult cartilage is avascular, immature cartilage can contain blood vessels as the cartilage is still undergoing edochondral ossification. Because of this, immature cartilage also appears thicker than mature cartilage, and cartilage continues to decrease in thickness as a person ages [143–145]. The cellularity, likewise, decreases with age [146]. Miotic cells can still be seen in immature cartilage. With the development of a defined calcified zone and, later, closure of the epiphyseal plate, cartilage division is seldom seen in healthy tissue. In addition to fewer chondrocytes, age also brings about lowered metabolic activity, increased apoptosis, and subdued response to growth factors [147–149], characteristics that are antithetical to healing and to the facile use of older chondrocytes in tissue engineering.

Collagen crosslinking has been observed to increase with age [150]. Increased glycation (non-enzymatic glycosylation, where sugars are added) of this matrix component also makes cartilage stiffer, but more brittle and prone to failure [151]. Younger cartilage displays greater birefringence, indicative of a greater degree of collagen alignment, as compared to older tissues [152]. Lastly, collagen type XI fragments are only seen in young cartilage (below nineteen years of age), a possible indication the collagen turnover slows down significantly beyond this point [153].

Significant changes are also seen with the proteoglycan content of cartilage with age. As the tissue matures and ages, proteoglycan content decreases with concomitant reductions in the protein core size, resulting in molecular weight decreases [154]. Chondroitin content decreases [155], and link protein fragments with time [156], reducing the amount of aggrecan in the tissue. Despite these changes, the permeability was not found to change much with age, and the equilibrium modulus value of the tissue only decreased slightly [157]. However, these properties appear to depend on when the data are taken, as others have shown cartilage to reach a peak in stiffness and elastic energies around forty years of age, while the viscoelastic energy reached a peak much earlier at 16-29 years, with steady declines of the above values over time [158].

Though the material properties of cartilage may not change much with age (or, as an aside, dietary restrictions) [159], repeated, non-physiological loading of the tissue can result in defects, as presented in the next section. Particularly problematic is the lack of a sufficient healing response displayed by these cells, a response that also diminishes with age.

It is also worthy to note that other systemic changes that occur with age can also affect articular cartilage properties. For instance, a study performed on ovariectomized sheep shows decreased articular cartilage material properties if left untreated with estrogen replacement [160], a result relevant to menopausal women. Diabetes, too, can affect cartilage properties adversely [161]. Anti-inflammatory treatments such as intra-articular administration of methylprednisolone has also been shown to decrease cartilage material properties [162]. All these factors can make the cartilage more prone to injuries.

2.3   CARTILAGE INJURIES

2.3.1   OSTEOCHONDRAL, CHONDRAL DEFECTS, AND MICROFRACTURES

Injurious impact and repeated loading, torsional loading, joint malalignment, and foreign bodies in the joint can all lead to cartilage injuries. Cartilage injuries are classified as osteochondral defects, chondral defects, and cartilage microfractures. Despite being avascular, the cartilage "bleeds" from an osteochondral defect because the injury extends through the cartilage into the subchondral bone. Chondral defects are also visible to the naked eye, oftentimes through India ink staining, in contrast to cartilage microfractures. For each type, the changes in tissue appearance, composition, and mechanical properties will be presented below along with a description of the resultant cellular responses. The information presented earlier on articular cartilage physiology and aging directly affect the outcomes for cartilage injuries, and injuries often lead eventually to osteoarthritis, described in the next section. For instance, a lack of vasculature and the hyaline nature of articular cartilage result in few cells available to mount an adequate healing response. A degenerative process thus follows, exacerbated by age, as stresses from daily use continue to be applied onto the already weakened tissue, with morphological and cellular changes akin to those seen with osteoarthritis.

The Outerbridge classification has been used to grade the severity of cartilage lesions. Arthroscopic application of this scale has shown that orthopaedic surgeons can use the Outerbridge classification to accurately grade chondral lesions regardless of their level of experience [163]. With this scale, Grade 0 represents normal cartilage [164]. Cartilage with softening and swelling is classified as Grade I. Grade II, which is the most frequently observed clinically [165], denotes that a defect contains fissures that do not extend to the subchondral bone (i.e., a chondral defect) and that the defect's diameter is less than 1.5 cm. When fissuring does extend to the level of the subchondral bone, in an area with a diameter more than 1.5 cm, the lesion is classified as Grade III. Lastly, articular cartilage injuries that result in exposure of the subchondral bone is classified as Grade IV [164]. Noyes and Stabler [166] developed a scale specifically for arthroscopic grading of cartilage. A similar scale modified from the Outerbridge scale [167] has been shown to correlate well with the Histological/Histochemical Grading System (HHGS) when the severity of the lesions was Grade III or below [168]. Several grading scales have also been developed for osteoarthritis, such as the Osteoarthritis Research Society International (OARSI) and Cartilage Histopathology Assessment System (OOCHAS) [169].

Cartilage microfractures do not result in immediate changes in the matrix that are visible to the naked eye. However, the damage to the collagen network begins to effect superficial GAG loss [170]. As the network loosens, increased hydration is also observed [171]. Microfractures can also lead to altered load distribution of the matrix, resulting in stress concentrations that can cause further damage or a greater proportion of forces borne by the bone. These loading alterations, as well as fractures to the calcified layer that can occur, lead to eventual thickening of the subchondral bone [172]. The calcified layer also thickens as the cartilage thins. Since cartilage is aneural, repeated loading of microfractured cartilage can continue without pain, leading to further degeneration [173]. Though the GAG loss stimulates chondrocyte activity, the metabolic response is typically inadequate, leading to a net loss of GAG, increased wear, and the eventual development of fissures.

Chondral fissures are defects that do not extend to the subchondral bone. These defects are visible to the naked eye, often via India ink staining. Chondral defects can proceed from cartilage microfractures or from trauma, improper loading, or foreign bodies. Without blood, the intrinsic metabolic activity after such an injury is insufficient to result in adequate repair, leading to the eventual development of osteochondral fissures [173]. Osteochondral fissures are lesions that cross the tidemark and penetrate the underlying bone. Though growth factors and progenitor cells are recruited from the bone's vasculature, there is impaired functionality of the repair tissue, a mix of fibro- and articular cartilages [174], resulting in eventual degeneration into osteoarthritis. It is important to recall that, with age, both chondrocytes and progenitor cells decrease in number and metabolic activity, thus contributing to the cartilage healing problem.

2.3.2   CAUSES OF CARTILAGE INJURIES

Cartilage injuries can result from impact and repeated loading, and these can occur under a wide range of loads, time scales, and frequencies. Determining thresholds that can correlate to certain elicited cellular responses is important in understanding cartilage injuries and degeneration. Because of the viscoelastic nature of articular cartilage, load rate affects tissue stiffness and thus failure. The rates of applied stress, strain, and load must be considered. For example, using a confined compression loading protocol, the dynamic modulus has been shown to increase from 225 to 850 MPa as the load rate was increased from 25 to 1000 MPa/s [175].

As reviewed elsewhere [176], peak forces during normal physiological loading of knee articular cartilage range from 1.9 to 7.2 times body weight. For a 70 kg person, this would correspond to ∼1400 to 4900 N [177]. Taking the medial tibial plateau to have an area of 1670 mm^2 [178], this corresponds to a maximum stress of ∼3 MPa. During normal activities, like running, time to peak force is on the order of 30 ms, leading to a stress rate of 100 MPa/s. One can expect that impact injuries occur on a time scale an order of magnitude smaller, resulting in stress rates of 1000 MPa/s. Based on data such as these, Aspden and associates [179] put forth a definition for injurious impact loading as time to peak load on the order of milliseconds plus one of the following: (1) stress rate greater than 1000 MPa/s, (2) strain rate greater than 500 s^{-1}, or (3) loading rates in excess of 100 kN/s.

Cartilage subjected to loads that do not satisfy Aspden's three criteria for impact can nonetheless result in damage via repeated application. The thresholds of forces and timescales required to produce injuries are important in determining the type of injury sustained and for modeling impact and injurious compression experimentally. Mathematical models of impact and injurious compression have been performed [7,180–183], as well as studies of impact done on explants [175,177,184–191], and *in vivo* [171,192–198] based on this and other classifications. The difference between impact and injurious compression is that the latter occurs over a longer time span [199,200]. Using a drop tower device to apply impact on cartilage [190], it was noted that, even at levels where no morphological changes were observable immediately, significant cell death and decreases in cartilage stiffness were found four weeks after impact, as compared to unimpacted controls [189]. These data show that "clinically silent" impacts can nonetheless result in articular injuries.

The superficial zone contains collagen fibers that are oriented parallel to the cartilage surface, aligned in the direction of shear [32]. One important function of this higher collagen and low proteoglycan zone is to resist shear stress, and the superficial zone has been shown to be stiffer in tension than the other zones [61]. In addition, cartilage is weaker in tension in the direction perpendicular to shear [64,66], and this weakness makes it susceptible to torsional injuries. Classifications of torsional injuries with respect to the forces, time frame, and frequencies necessary to generate cartilage microfracture, chondral, or osteochondral defects are not as well-studied as impact and injurious compression loading. These types of loading, are, however, likely to play a role in cases of joint malalignment, as the altered biomechanics can result in repeated non-physiological focal stresses or torsion.

Surgical procedures or injuries to other connective tissues can be a cause of joint malalignment or improper cartilage loading. Also, foreign bodies, such as crystals, can result in damage to the articulating surface. Monosodium urate and calcium pyrophosphate dehydrate [201], characteristic in gout and synovitis [202,203], are strongly associated with cartilage lesions, higher levels of superficial zone protein and collagen X [204], and are suspected to be linked to arthritis [205]. Associated lesions appear to be biomechanically induced, and crystals have been observed in joints both before and after the onset of OA [204]. Plastics or other debris can also contribute to third body wear in a joint [206].

2.3.3   REPAIR RESPONSES TO CARTILAGE INJURY

Cellular responses to osteochondral defects are complex due to the involvement of cells both from the articular cartilage and elsewhere. For cartilage microfractures and chondral defects, the dense matrix keeps other cell types out of the repair response. Radiolabeling and other techniques have shown chondrocyte proliferation and matrix synthesis occurring for about two weeks post-injury [207,208]. However, before the defect is filled, this anabolic activity ceases and the matrix is left unprepared for the continued rigors of daily use, leading to eventual degeneration as described previously [209,210].

With osteochondral defects, the repair process extends greatly beyond two weeks. The sub-chondral vasculature delivers progenitor cells that are much more swift and active than chondrocytes, dominating the healing process. Blood from the bone forms a fibrin clot, which contains platelets that secrete factors to recruit mesenchymal stem cells from the bone. In the two weeks following injury, MSCs proliferate and differentiate. Repair continues beyond two weeks as differentiated cells produce collagen type II and collagen type I. By six to eight weeks, the defect is filled [174,211,212]. From here on, matrix production slowly shifts from collagen type II to collagen I such that by the end of one year the repair tissue consists of both hyaline cartilage and fibrocartilage [207,210,211]. Since the fibrocartilage does not possess sufficient mechanical properties for sustained function, continued matrix degeneration via fibrillation [213], GAG loss [214], chondrocyte death and proliferation, and development of deep fissures are observed beyond the first year [174,210]. Repair responses from both chondrocytes or other cells are also seen with immature cartilage where calcification has yet to be completed and where vasculature still exists close to the articulating surface. However, it has been shown that cartilage defects do not heal even in immature animals [208,209]. Whereas it has been proposed that "non-critical" defects (under 3-9 mm in diameter, depending on the animal model) can heal [215], the repair tissue is again fibrocartilage, which eventually degenerates, leading to osteoarthritis.

2.3.4   COSTS OF ARTICULAR CARTILAGE INJURIES

In a retrospective cross-sectional study, the follow up costs for the first five years following arthroscopy and treatment for 1,708 Germans between 1997 and 2001 was quantified. The treatments included mostly debridement/cartilage shaving, with abrasion arthroplasty, chondroplasty/laser chondroplasty, and microfracture or subchondral drilling performed at roughly the same frequency. Autologous chondrocyte implantation, osteochondral allografts, and autografts were also observed, though much less frequently. Not included in the study were cases that were grade 4 according to the Outerbridge classification system, osteoarthritic (Fairbanks greater than grade 3), or consisting of bacterial infections or tumors. Cumulative costs associated with loss of productivity were found to be almost four times that of the direct costs, with those who had prior operative history on the knee spending roughly double [216]. Another source of traumatic injuries posing significant costs is combat trauma. Based on queries to the Department of Defense Medical Metrics (M2) database for the hospital admissions and billing data between October 2001 and January 2005 for injuries sustained in Iraq and Afghanistan, estimates to the cost of combat-related joint injuries to approach

$2 billion [217]. Without a sufficient healing response, cartilage lesions eventually degenerate to osteoarthritis, which has much greater associated costs (in excess of $65 billion [218]) as discussed in the next section.

2.4   OSTEOARTHRITIS

Osteoarthritis is a significant problem in the US, especially with the aging population, as it poses great costs both financially and to the patient's quality of life. According to the National Health and Nutrition Examination Survey I (NHANES I), 12.1% of the US population ages 25-74 years had clinically defined OA of some joint [219]. With the aging population, the incidences of OA have increased from an estimated 21 million in 1995 to nearly 27 million a decade later [220]. Broken down by age, surveys show that in the US, over a fifth of the population over 45 and almost half over 65 develop OA [221]. OA is clinically divided into two types: primary (cases with no known cause) and secondary (cases with identifiable cause). Known as post-traumatic OA, a majority of secondary OA is attributable to traumatic joint injury that may have occurred years previously [222–224]. Mechanical injuries, such as those that happen during motor vehicle collisions, falls, and sports injuries, have been implicated in the development of post-traumatic OA, though the precise pathophysiology is not yet fully understood [225–227]. Diagnosis of OA via radiographic evidence (based on the presence of osteophytes) shows that the incidence of OA in people over the age of 45 is 27.8% in the knees and 27.0% in the hips [220]. Symptomatic OA manifests itself as frequent pain in a joint and radiographic evidence of OA in the same joint. It is important to note, though that the pain may not be from the arthritis seen in the joint [220]. In addition to costs associated with treating the articular surface, other medical comorbidities observed with OA and rheumatoid arthritis (RA) contribute significantly to the costs of disease management [228].

While replacement cartilage for OA will address a significant clinical problem, tissue engineering may be used to repair focal defects before such lesions manifest themselves into OA. The aforementioned statistics on cartilage defects indicate that the majority are observed on the patella or the medial femoral condyle [165,229]. The altered biomechanical environment resulting from ACL transection has long been used as a model to induce osteoarthritis. Contributive to the degeneration is the increased knee-abduction moment post-ACL transection [230]. Meniscus injuries have been linked to osteoarthritis and meniscectomy has been linked to both increased prevalence and severity of cartilage injuries [231]. An established experimental method to induce OA via meniscectomy has shown that the patellar cartilage is adversely affected within three months in an ovine model [232]. Clinical studies of patients with meniscectomy have shown that, after twenty one years, mild radiographic changes were found in 71% of the knees, while more advanced changes were seen in 48% [233].

As described above, causes of OA have been categorized into primary and secondary. Trauma was first linked to OA by Hunter in 1743 [234]. Athletes with a history of joint injury have higher incidence of OA than their peers [235]. ACL injury causes immediate changes in biomechanics [236–238], which may lead to OA [239,240]. For instance, the quadriceps and surrounding muscles of the knee are prevented from full activation in injured knees, and this arthrogenic inhibition has been observed in patients with ACL and other joint injuries [241–244]. It has been postulated that the resulting alteration in biomechanics is partially responsible for OA development due to the muscles' role in energy absorption [245]. The altered loading not only leads to possible lesions [246, 247], but also to different compressive loads or loading rates applied to chondrocytes that may lead to catabolism [248–250]. Shear has also been linked with chondrocyte death and secretion of proinflammatory factors [251,252].

Fissure propagation is affected by cartilage thickness and the ratio between the stiffness of the cartilage and its underlying bone [253]. Modeling has shown that higher stresses are found in the thicker of two contacting biphasic layers of articular cartilage [254]. Considering that OA both reduces cartilage stiffness and thickness, these models are evidence that the disease results in a degenerative cycle. More recently, a three-phase (collagen, matrix, and synovial fluid), transversely isotropic, unconfined half-space model of articular cartilage was created for studying surface fissures [255]. Interestingly, it was shown that collagen is in tension in the first 10 to 20 s of a rapidly applied compressive load, switching to compression thereafter if the load is held constant. The tensile stresses generated were within the range of reported tensile strength of collagen fibers, showing that failure of collagen could lead to surface fissuring. This was not the case for slowly applied loads. In another study [256], mediators of collagen damage due to mechanical injury were investigated in a fibril-reinforced poroviscoelastic model of articular cartilage. Using differential immunohistochemistry, wherein distinct antibodies were used to separate staining of enzymatically cleaved collagen from other damaged collagen, it was shown that shear and maximum strain in the collagen fibers corresponded to areas staining for mechanically damaged collagen. The results from investigations of surface fissuring implicate collagen as having a key role in keeping the surface intact when subjected to injurious mechanical loading.

Whereas abundant evidence exists to indicate that osteochondral injuries lead eventually to OA, the cases to justify chondral defects and microfractures as similar causes are not as clear cut. As pointed out earlier, though, when cartilage is subjected to impact, degenerative changes can occur even when no noticeable signs of damage can be observed immediately post-impact [189]. Investigations have been made to see if agents such as P188 [177, 257, 258], IGF-I [177], and doxycycline [191] can halt or reverse the degenerative process post-impact, and these will be discussed more extensively in Section 3.4. As investigators continue to uncover the biochemical and biomechanical pathways that lead to OA, no established methods currently exist in stopping these degenerative changes, leading to significant costs both financially and to the quality of life of OA patients.

2.4.1   OSTEOARTHRITIC CHANGES IN THE MATRIX

Grossly, OA affects all joint components. Starting with the changes associated with the capsule, thickening and frequent adherence to the underlying bone is seen, with increased vascularization and hemorrhage [259]. Amyloid formation is often observed with advanced stages of the disease [260,

261]. The subchondral bone remodels with thickening as it is now unprotected from the load borne by cartilage [262, 263]. The bone may be entirely exposed as cartilage is denuded during the disease's progression. Along with the thickening, new bone formation is observed. The remodeling and formation of osteophytes, which arise from the bony margins of the joint, alter the contours of the joint. Osteophytes are strongly associated with malalignment [264–266], an identified cause for cartilage lesions that can lead to further degeneration. Depending on the joint location, the cartilage can be white to yellow or brownish, with generally decreased mechanical properties, though there can be regions where the new, healthy-looking cartilage has formed in small, pebbled patterns.

Destruction of the cartilage surface in OA occurs in phases. During the development of the disease, no visual, functional, or mechanical alterations appear detectable. Fibrillation, surface erosion, and fissures are the first noticeable signs of the disease. From this point on, altered histological staining will show continued decreases in proteoglycan content. The tidemark begins to appear irregular, punctuated with blood vessels. A second stage of the disease shows greater surface wear and irregularity. Vertical and sometimes horizontal fissures can be seen in the cartilage. Proteoglycans start to leave from the fissures and an absence of staining will spread from these areas to the rest of the tissue. These patterns of increased tissue fragmentation and decreased staining continues until the cartilage, completely robbed of its abilities to withstand load, is worn away to expose the subchondral bone. During these cartilage changes, bone and synovium remodeling occurs as described previously.

When considering this cascade of events, it is not surprising to note that OA cartilage possesses inferior mechanical properties. The gradual proteoglycan loss is a first hint. However, OA cartilage also increases in water content, an observation that appears counter-intuitive, as it is the negative charges on the proteoglycans that attract ions to increase the osmotic pressure that drives hydration in this tissue. Lowered proteoglycans are thus expected to lower the Donnan osmotic pressure and result in water loss. It is postulated that the observed increase in hydration is due to the loosening of the proteoglycan network to allow for macromolecular un-curling, thus allowing for more interstitial space for water to occupy. This theory is supported by the observations that the fraction of aggregating proteoglycans decrease with OA and that the proteoglycans are increasingly more extractable with disease progression. Being unable to aggregate, the proteoglycans are thus unable to pack as densely, supporting the uncurling hypothesis. Greater extractability is likely a result of decreased molecular weight or, as has been shown, the result of damaged link proteins that no longer facilitate the formation of aggrecan. In addition, the catabolic agents released during OA can loosen the collagen network, too, to result in more space for water to occupy. Tensile strength, attributed to the collagen in cartilage, has been shown to decrease with OA.

2.4.2 PROLIFERATION, CATABOLISM, AND CELL DEATH

The cellular behavior in OA can be classified into three stages. With changes occurring in the interterritorial and territorial matrices, the chondrons, where the chondrocytes reside, may become swollen and distended, signaling the cells to proliferate to fill up the additional space. IL-1 upregulation then initiates the destruction of the fibrillar collagen environment, which results in additional chondron distortion, and proliferation continues. Collagens type II, IX, and XI are rendered more soluble by the catabolic activity. The breakup of the collagen network results in TGF-β being released from the matrix, and greater collagen type VI production and sequestration by the cells ensue. Finally, many cells, encapsulated by a sheath of collagen type VI, now occupy a distorted chondron, where one or two cells used to reside [267].

Cartilage has long been considered immune-privileged because it is avascular and alymphatic. The dense, hyaline matrix greatly hinders cell movement and, as discussed earlier, is a barrier to even chondrocytes themselves, precluding a complete healing response as the sequestered chondrocytes are prevented from populating the wound edge. This same dense matrix also serves as a barrier to integration, which will be discussed in the last section of this book, in Section 5.2.2. For the time being, it is important to note that, unless compromised, this dense tissue locks cells and matrix in, and keeps cells out. How, then, does inflammation wreak such havoc on this tissue? As the disease progresses the thickening and increasingly hypervascular synovium secretes matrix metalloproteinases (MMPs) and aggrecanases. In addition, chondrocyte death can be observed much earlier than matrix degradation, and as a result, releases secondary necrosis factors that diffuse to initiate apoptosis in neighboring cells. In this case, the dense matrix becomes a space that the chondrocytes cannot escape from, and the chondron a place where phagocytes cannot reach to clear apoptotic debris to interrupt the cycle of inflammation and necrosis [268].

Other molecular factors that appear to serve as players in OA progression include F-spondin, a neuronal extracellular matrix glycoprotein. Approximately a 7-fold increase for this species is seen in OA cartilage, and its presence primes TGF-β1 and prostaglandin E2 (PGE2) release [269]. PGE2 is a proinflammatory mediator, and its release is also accompanied by collagen degradation and MMP-13 activation when cartilage explants are stimulated with F-spondin. The increased synthesis of COMP by chondrocytes and synoviocytes has also been associated with OA, as its production can be stimulated by TGF-β1 [270].

2.4.3   COSTS OF ARTHRITIS

OA causes significant pain and suffering to individual patients, and the economic burden of this disease for society is great [271]. For example, in the US alone OA related costs exceed $65 billion per year (both medical costs and lost wages) [218]. Conservatively, it is estimated that 1 in 8 American adults over the age of 25 have clinically manifested OA [220, 272], making it one of the leading causes of disability in the United States [225]. Increased cost associated with OA and rheumatoid arthritis (RA) are high as it has been demonstrated that several other comorbid conditions exist, such as anemia, osteoporosis, and bacterial infection [228]. Analyses show that, when compared to patients without OA and RA, significantly more charges are incurred by OA and RA patients in other areas such as respiratory, cardiovascular, gastrointestinal, neurological, and psychiatric conditions, and also for general medical care. Increased therapeutic procedures, physician services, use of prescription medication, etc., are also more prevalent in sufferers of OA and RA [273]. Another significant source of costs is loss of productivity. Indirect costs associated with articular cartilage injuries (that have not yet progressed to OA) can be four times as much as treatment of the defect itself [216], and a similar scenario can be expected for OA.

2.5   MOTIVATION FOR TISSUE ENGINEERING

Based on over 25,000 arthroscopies surveyed, it has been shown that osteochondral and chondral lesions are the most common, accounting for 67% of the observations, while OA accounts for 29% [165]. As presented earlier, the cartilage's inability to mount a sufficient healing response eventually results in degenerative changes, and the proportion of lesions to OA observed is expected to change in the near future due to the baby boomers, with concomitant rises in management and treatment costs. Aside from OA being linked to the aging population, it is more important to note that these lesions frequently occur in the youth, a population whose needs for long-term solutions are much greater than their elders.

A need for tissue engineering rises from the prevalence of joint injuries in adolescents. "Little Leaguer's Elbow," osteochondrosis, and osteochondritis dissecans are joint diseases that occur mainly in children, due to the increased vulnerability to stress in the growing skeleton. One out of three school-age children will sustain an injury severe enough to require medical treatment. Emergency room visits are the highest among children and young adults. ACL treatments, as well, are seen in higher frequencies (and are rising) in these two groups [274, 275]. With the estimated 30 million children who participate in organized sports activities, the yearly costs for injuries within this group have been projected to be $1.8 billion [276]. Kids may play on multiple teams with overlapping schedules, and it is not uncommon to see the absence of well-defined standards for when training becomes excessive. Little League began to implement a pilot pitch-count program only in 2005 [277], though its initially set standards were quickly relaxed. While elbow and shoulder injuries are common in baseball, a global survey of adolescent knee injuries put the incidence rate at greater than 25% in sports participants [278], particularly in basketball [279, 280]. In terms of articular cartilage defects, young patients with knee injuries show 75% superficial (grade I–II) and 25% deep lesions (grade III–IV) [229]. This is particularly alarming as data from 1995, 1996, and 1997 indicate that roughly 20% of the knee injuries in adolescents required surgery [281–283]. The urge to succeed that comes from the child, parents, and coaches has gone to such a degree that overuse injuries are common. It is difficult for some parents to realize that their children's hard work in sports can result in catastrophic cartilage injuries. Unfortunately, no consistently successful solutions exist for the cartilage repair problem in children and adolescents.

The formation of repair fibrocartilage serves only as a temporary biomechanical fix, and a long-term solution for youth afflicted with joint injuries would be ideal. Even in cases where the articular cartilage is not damaged in the primary traumatic event (e.g., osteophytes and ligament damage), the malalignment that can result [284] has been shown to predict cartilage loss [285]. A survey of global adolescent knee injuries shows that females are more prone to these injuries than males. Recent estimates put the rate of incidences at greater than 25% in sports participants [278]. Basketball has been linked to the highest rates of knee injury, as the frequent jumping associated results in loads several times body weight to be applied [279, 280]. Indeed, knee and hip cartilages show greater incidences of lesions than other anatomical regions. A survey of all cartilage lesions across all ages shows that the patellar articular surface and the medial femoral condyle were the most frequently damaged, accounting for 36% and 34%, respectively, of the cases surveyed [165]. These joints are also the ones that enable mobility and have a great impact on the quality of life. Arthroscopic evaluation in young Finnish males showed that 73.5% of the lesions were patellar, 12.0% in the medial condyle of the femur, and 8.0% in the femoral groove. Roughly, 75% of the patients had superficial (grade I–II) and 25% deep lesions (grade III–IV) [229]. This is particularly alarming as data from 1995, 1996, and 1997 indicate that roughly 20% of the knee injuries in adolescents required surgery [281–283]. Unfortunately, current therapies, as reviewed in the last section of this book, are not sufficient in effecting long term relief and activity resumption. Follow up studies of 5, 10, and greater years have consistently shown a need for improvement in the outcomes of arthroplasty, osteochondral, and autologous cell transplantation. Effective solutions are clearly needed, and these can be improvements on current therapies, chondrocyte transplantation, which can be considered as a form of *in vivo* tissue engineering, or controlled manipulation of cells and materials *in vitro* to form implantable neocartilage.

2.6   CHAPTER CONCEPTS

- The cartilages of the limbs mostly form by mesenchymal condensation, proliferation, and differentiation.

- A variety of chemical signals, such as TGF-β, BMP, VEGF, shh, etc., regulate the process of cartilage formation, and these signals have been manipulated to study and to recapitulate this process.

- Immature cartilage can contain vasculature. As cartilage ages, the vascular regions calcify, and cartilage thins.

- Aging results in increased collagen crosslinking, lowered collagen alignment, and slower collagen turnover. Proteoglycans also decrease in amount and size with age.

- Hormones and steroids can negatively affect cartilage material properties.

- Cartilage can be injured by impact, repeated loading, torsional loading, joint malalignment, and foreign bodies in the joint space.

- Sometimes, injuries to the cartilage show no gross morphological changes, and chondrocytes do not respond adversely to the insult immediately. However, chondrocyte death and catabolism have been shown to occur even for these "clinically silent" injuries.

- Chondral lesions do not heal, and osteochondral lesions are filled with mechanically inferior fibrocartilage that breaks down with use.

- The costs of cartilage injuries can be significant to both the young and the elderly. Traumatic injuries are seen with automobile accidents and, recently, with combat trauma.

- Osteoarthritis (OA) affects over one fifth of the US population over 45 and almost one half of those over 65.

- OA cartilage loses proteoglycans and possesses lower mechanical properties. Wear of the cartilage can lead to complete destruction of the articular surface and significant pain.

- In the US alone, costs of OA are in excess of $65 billion per year (both medical costs and lost wages). Comorbidities are common with OA and are also costly to manage.

- In addition to the elderly, cartilage injuries in children and adolescents are increasingly observed, with roughly 20% of knee injuries in adolescents requiring surgery.

CHAPTER 3

In Vitro Tissue Engineering of Hyaline Articular Cartilage

For decades, hyaline articular cartilage has been a primary target for tissue engineering efforts due to the lack of functional regeneration within the joint. In addition to focal defects, systematic problems such as OA can destroy the entire cartilage surface, resulting in loss of function and persistent pain. This chapter highlights both the seminal tissue engineering studies focused on hyaline cartilage, as well as the latest approaches that incorporate bioreactors, bioactive molecules, and specialized biomaterials.

Tissue engineering, in its classical sense, involves the manipulation of a complex interplay among biomaterials, growth factors, and cell populations [286] to achieve functional improvement or restoration. Articular cartilage has been a high priority for tissue engineers since it does not naturally regenerate after injury. Furthermore, the annual health care costs associated with musculoskeletal diseases and injuries are extremely large, and an effective reparative solution would not only reduce costs but also improve the quality of life for millions [287]. The average age for patients undergoing arthroscopy that exhibit cartilage defects in the knee is 43, and, combined with the demographical data on adolescent cartilage injuries as discussed previously, the need to create a repair tissue that can last several decades is a major goal [288]. The earliest attempts at cartilage regeneration involved transplanting either minced cartilage tissue or dissociated chondrocytes [289]. Surgical solutions to cartilage defects typically include surface abrasion, microfracture, and debridement, which all can reduce symptoms. However, the repair tissue formed in response to these procedures is fibrocartilage, which has biomechanical properties that are markedly different from normal cartilage [288]. Fibro-cartilage does not have the biochemical composition or structural organization to provide proper mechanical function within the joint environment and will degrade over time because of insufficient load-bearing capacity [290, 291]. Because of this, current research is striving to produce a tissue that is hyaline-like in its biochemical composition and mechanical properties. The first section in this chapter will focus on *in vitro* tissue engineering approaches. Attempts to tissue engineer within the *in vivo* environment will be discussed along with germane immunological considerations, as presented in the last chapter of this book.

Early on, it was thought that the *in vivo* environment should contain all the conditions necessary to effect successful regeneration. That is, the *in vivo* environment contains the proper growth factors and mechanical stimuli, delivered in a well-sequenced manner through autocrine and paracrine signaling, to effect proper healing, the major missing component being metabolically active chondrocytes at the defect site. Initial efforts at delivering mechanical stimuli *in vitro* attempted to emulate these signals. It has been described that the objective of bioreactors is to create signals reminiscent of the native environment, e.g., the 1 Hz pace of walking, the low oxygen tension of the joint, and others. Unfortunately, the physiological conditions have been shown repeatedly to result in cartilage degeneration, and, thus, the act of mimicking these conditions is now being questioned. It may not be that non-physiological conditions are required, just that physiological conditions of a different developmental period may be more beneficial in generating functional cartilage. To investigate this latter case, *in vitro* tissue engineering has been employed to recapitulate developmental conditions, in contrast to the *in vivo* tissue engineering efforts, which can only apply adult conditions akin to the healing response.

3.1   THE NEED FOR *IN VITRO* TISSUE ENGINEERING

The primary advantage of *in vitro* tissue engineering is proposed to be immediate functionality. A tissue replacement that is mechanically and biologically functional before implantation will have a higher probability for success. This is especially true for mechanically rigorous environments such as articulating joints. Without the requisite mechanical characteristics, a tissue engineered construct would be quickly destroyed by the normal loading of an ambulatory patient. However, the implantation of a construct that possesses material properties comparable to the native tissue would not fracture or degrade. Because of this, many researchers believe articular cartilage engineering should place emphasis on construct development *in vitro*. Since the tissue resides in a mechanically demanding environment, the implanted construct needs to be developed to a point that it can withstand or respond to these mechanical loads. Constructs possessing insufficient integrity will collapse in the articular defect, which not only prevents regeneration but could also accelerate degradation of the tissues surrounding it. Efforts to heal large defects *in vivo* could fail without some means of protecting the structure of newly developed tissues. By growing neocartilage in a laboratory, the culture environment can be carefully controlled with respect to nutrient supply, biological stimuli, and mechanical loading.

For a tissue like articular cartilage, possible treatments often depend on the type of damage to the joint. For example, an osteochondral defect which reaches down into the subchondral bone introduces blood into the system. This influx of blood and marrow brings a variety of chemicals and cells to the injury site. However, fibrocartilage will form in the defects if left untreated, filling the site with a disordered mass of fibrous tissue that possesses no long-term mechanical functionality. Another type of damage in cartilage is termed a chondral defect and does not extend through the depth of the tissue to reach vascularity. In this case, some of the chemical and biological variables associated with osteochondral defects are not relevant. Unfortunately, the mechanical functionality is still compromised due to disruption of the tissue's surface. Both osteochondral and chondral defects can be considered focal defects since the damage is localized to a single region. The most difficult type of cartilage injury to treat is a systemic breakdown of the articulating surface caused by diseases such as OA and osteochondritis dissecans. Traditional tissue engineering approaches create small constructs that can be fit into focal defect sites in the cartilage. However, this would be insufficient for injuries affecting the entire joint since there would be no functional tissue to anchor the new constructs in. As yet, there are no successful approaches to treating OA using conventional tissue engineering. Researchers are continually investigating alternative approaches, such as engineering a replacement tissue that can completely resurface the joint [292]. Other possibilities include gene therapy or pharmaceuticals, which might have more success in treating systemic degeneration of articular cartilage.

Cartilage growth and development are affected by both biological and biomechanical stimuli. On the mechanical side, loading is a required part of the normal joint environment. As seen in previous sections, while excessive forces can damage cartilage tissue, some stimulation is necessary to promote chondrogenesis [4]. Articular cartilage will atrophy in a mechanically static environment [293], so researchers are currently evaluating a variety of loading approaches to prevent this while promoting the regeneration process. An important factor to consider prior to mechanical loading is the choice of scaffold used for the engineered construct. The scaffold material not only affects how cells sense mechanical loads, but also provides an environment that can influence cell attachment and matrix synthesis. In addition to mechanical stimuli, articular cartilage responds dramatically to growth factors that are naturally present in the joint environment. The TGF-β superfamily includes growth factors that are present in developing bone and cartilage. These molecules play an integral role in the natural development process, and *in vitro*, can induce dramatic effects on the growth of orthopaedic tissues. This section will illustrate the importance of the *in vitro* culture environment on the growth, development, and functionality of native and engineered articular cartilage. Following the accepted paradigm for functional tissue engineering, four main categories will be reviewed: cell sources, biomaterials, bioactive molecules, and bioreactors [294].

3.2   CELL SOURCE

Cells are one of the key components of tissue engineering. While an exogenous cell source is not absolutely necessary, studies have shown that including cells in an engineered construct accelerates regeneration *in vitro* and *in vivo* [295]. Furthermore, these implanted cells have been shown to remain in the tissue without being replaced by host cells [295]. Researchers have several options when choosing a cell source. The cell type most commonly used in early cartilage engineering studies is the autologous chondrocyte. By extracting cells from the patient's own body, any immune response is minimized or totally removed. Furthermore, chondrocytes already are differentiated into the target phenotype and have the capacity to secrete cartilage-appropriate matrix molecules.

The choice of cell type often depends on the initial condition of the cartilage tissue. In cases of extensive degradation or disease, use of autologous chondrocytes is not an option. One possible alternative is to use allogeneic chondrocytes from donor tissue. This approach is commonly used for general *in vitro* experiments and some *in vivo* studies due to the ready availability of donor tissue. While the cell phenotype is appropriate for the implant environment, problems can arise with respect to tissue availability for humans, as well as possible disease transmission or immune response.

Due to limits on the availability of human tissue, some have suggested that cross-species cell implantations might be an alternative option. Xenogeneic transplants have been successfully used in sheep [296], goats [297], and rabbits [298]. However, similar difficulties exist with xenogeneic transplants as with allogeneic transplants, namely immunogenicity concerns. Further complications could arise with cross-species compatibility issues at the cellular and molecular level.

A promising cell source for cartilage tissue engineering is autologous progenitor/stem cell populations [299]. Adult progenitor cells reside throughout the body and can be differentiated along many different lineages. Progenitor cells from bone marrow and fat tissue have been extensively investigated for their promising application to cartilage regeneration [300]. Also, dermis-derived cells exhibit significant promise [301]. If autologous cells are used, minimal problems with immunogenicity exist. Progenitor/stem cells show a large capacity for proliferation, so only small samples are needed to obtain enough cells to grow the large populations necessary for tissue engineering. Donor site morbidity and patient pain are dependent on the site of harvest, but for some cell types this is minimal (i.e., adipose-derived stem cells).

Another possible cell source akin to progenitor populations is embryonic stem cells. While progenitor cells can proliferate extensively, extensive expansion in monolayer culture can have negative effects on proliferative rates, telomere shortening, and loss of multipotency [302–306]. Embryonic stem cells, however, have an unlimited capacity for proliferation, and hence, are attractive for tissue engineering endeavors that require large cell numbers [286]. These cells are truly pluripotent, showing a capacity to differentiate into any cell type in the body. However, researchers do not currently know the best ways to differentiate embryonic stem cells along every lineage. Some protocols are more defined than others, though, and good results have been obtained for the chondrocytic lineage [299, 307–310]. As with all cell therapies using embryonic stem cells, there are potential problems with teratoma formation, poorly controlled cell proliferation or differentiation, and possible immunogenicity problems since the cells come from an allogeneic source. A more detailed description of the use of these alternative cell sources is provided in the Future Directions section.

3.3   SCAFFOLD DESIGN

For functional tissue engineering, biocompatible scaffolds are chosen to best fulfill a role in improving the regeneration of a damaged or diseased tissue. While recent studies have indicated that cartilage constructs can be formed *in vitro* using only cells [311–314], traditional tissue engineering approaches have seeded cells on scaffolds to provide structure to the neocartilage. The architectural structure of the scaffold can affect the mechanical properties of the construct, cell seeding distributions, and diffusional characteristics. Furthermore, the material itself can help or hinder cell attachment, proliferation, and synthesis over the lifetime of the implant.

The base scaffold material can be considered the central component of a tissue engineered implant. The scaffold should fulfill three main requirements: 1) have an interconnected network that allows efficient diffusion of nutrients and wastes; 2) be biocompatible and bioresorbable, with a degradation rate that ideally matches the rate of tissue growth; 3) and allow for cell attachment, proliferation, and differentiation. The last requirement is often fulfilled using bioactive molecules that are either physically tethered to the scaffold or included in the culture media as a biochemical stimulant.

An additional factor for choosing a scaffold is whether it will be used *in vitro* or *in vivo*. If implanted immediately, the scaffold should possess mechanical characteristics that are appropriate for the loading environment. The scaffold should maintain its shape and protect the seeded cells from any excessive forces. The degradation of the scaffold should correspond with the growth of tissue in the construct, which would gradually take more of the applied load from the scaffold. If chemical initiators are used to cross-link the scaffold either *in vitro* or *in vivo*, then the process should be designed to minimize negative effects on cell viability or metabolism. Some injectable biomaterials are crosslinked in the defect site to achieve sufficient mechanical properties, but this process can involve chemicals that are cytotoxic. Research is ongoing to find an injectable cell/polymer solution that forms a construct with robust mechanical properties in the defect site [315–317]. Cell-seeded scaffolds cultured *in vitro*, however, do not need the same level of structural integrity since newly formed tissue should help achieve mechanical characteristics sufficient for the biological environment. These properties would be independent of the degrading scaffold material. This approach simply uses the scaffold as a structure that helps support seeded cells for a period of weeks while new tissue forms. Ideally, once the construct is implanted, the newly formed matrix is developed enough to function successfully in the native loading environment.

For the purpose of this section, biomaterials are sorted into three main categories: natural polymers, synthetic polymers, and composites. Natural polymers are found in living organisms and can be extracted and processed into functional biomaterials. Synthetic polymers are created using chemical processes, which allow extensive customization of material properties. However, some processes can also have negative side-effects such as cytotoxicity or immune response activation. Composite scaffolds combine two or more materials into one scaffold to take advantage of special characteristics intrinsic to each substance.

New materials for biological applications are frequently, synthesized but extensive chemical and physical characterization are necessary before a material can be used in the body. The biomaterials summarized in this section have been well characterized and shown to be cyto-compatible in many, individual tissue engineering studies. Not all have been applied to cartilage tissue engineering, though, and success could be dependent on the cell-surface interactions specific to different cell types.

3.3.1 NATURAL SCAFFOLDS

Natural materials are often preferred for biological applications because they are believed to elicit little or no immune response. Their structures can vary from hydrogels (a colloidal gel in which water is the dispersion medium) to solid fibers and fragments. Among the natural materials used in cartilage engineering are alginate, agarose, chitosan, fibrin glue, type I and II collagen, hyaluronic acid-based materials, and reconstituted tissue matrices. Each material has strengths and weaknesses to its use, and results can vary depending on the application.

Alginate is a polysaccharide extracted from algae and can be used to encapsulate cells in a three-dimensional matrix. Encapsulation maintains a chondrocyte's rounded morphology, which has been shown to induce re-differentiation of monolayer expanded cells [318]. This approach can also be applied when differentiating stem cells along the chondrocytic lineage. Besides encapsulation, one of the main advantages of alginate is its proven biocompatibility [319]. For sterile applications, alginate can be purified by filtration, precipitation, or extraction. However, alginate is not an ideal material for many tissue engineering applications. The material does not degrade rapidly *in vivo*, which can interfere with new tissue growth. Long-term implants encounter problems with alginate since the scaffold loses its integrity within a year [319].

Agarose is a polysaccharide derived from seaweed that exhibits temperature-sensitive solubility in water, an attribute convenient for encapsulating cells [319]. Similar to alginate, agarose provides a biocompatible, three-dimensional environment for culturing chondrocytes. Unfortunately, the degradation properties of agarose are similar to alginate and cannot be easily altered to tailor the life of the scaffold. In addition, it is unclear whether agarose is eventually degraded or removed as the cells make matrix. Despite these deficiencies, many *in vitro* studies use agarose as a scaffold material when investigating the effects of mechanical stimuli on chondrocytes [320, 321]. Since it is a continuous, hydrogel matrix, applied mechanical forces are transmitted to the embedded chondrocytes, stimulating them to produce extracellular matrix proteins [322].

Another common scaffold material used for cartilage tissue engineering is collagen, specifically collagen type I or II. Collagen is the major component of extracellular matrix in connective tissues. As with most other natural materials, collagen has to be processed before use to decrease its antigenicity. Collagen type I scaffolds have facilitated cartilaginous tissue formation in studies investigating direct compression [323] and cross-linked proteoglycans [324, 325]. However, this material alone can also result in dedifferentiation of chondrocytes [319], likely due to the fact that type II collagen, not type I, is the predominant collagen in native articular cartilage. Cells seeded onto type II collagen scaffolds show a retention of the chondrocytic phenotype [326]. Unfortunately, fabricating collagen type II scaffolds is a difficult and expensive process compared to other natural materials due to its limited availability.

Chitin is a semi-crystalline polymer derived from the exoskeleton of crustaceans. After deacetylation, chitin is termed chitosan and is a natural biomaterial possessing a high degree of biocompatibility *in vivo* [319]. The molecular structure of chitosan is similar to many glycosamino-glycans, allowing it to interact with growth factors and adhesion proteins [319]. The degradation of chitosan is controlled by the degree of deacetylation within the polymer, which can be altered during processing of the original chitin material. Unlike the natural materials described previously, chitosan scaffolds can degrade rapidly *in vivo* to allow space for the formation of new tissue [319]. The porosity of the biomaterial can also be controlled during processing, effectively modulating the overall strength and elasticity of the scaffold [327]. Oftentimes, chitosan is combined with other molecules to create scaffolds that stimulate the secretion of cartilage matrix. One such study cross-linked chondroitin sulfate with chitosan to form a scaffold that promoted the chondrocytic phenotype [328]. Chitosan and chitosan composites can also influence cell attachment and growth in culture. Endothelial and smooth muscle cells seeded onto a dextran sulfate-chitosan composite, heparin-chitosan composite, or chitosan material alone all showed positive effects on cell attachment and proliferation. However, a GAG-chitosan composite actually inhibited attachment and growth [329]. The modification of chitosan scaffolds with proteoglycans can dramatically change the overall characteristics of the scaffold. This flexibility is an attractive attribute that could make chitosan a beneficial material for articular cartilage engineering.

Silk is a naturally occurring polymer that is extruded from insects or worms and has been increasingly used in biomedical applications. The material has good biocompatibility, slow degradation rates, strong mechanical strength, and can be processed into many different forms useful for tissue engineering [330]. Recent studies investigating silk scaffolds for cartilage engineering have shown good results for chondrogenesis in seeded stem cells. In comparison to collagen-based scaffolds, silk constructs had higher type II collagen and GAG deposition, as well as better chondrocytic gene expressions in seeded mesenchymal stem cells [331].

Fibrin glue is another naturally-derived biomaterial that has been used in tissue regeneration therapies. It is made by mixing fibrinogen with thrombin, which acts to solidify the material either in a defect site or in another scaffold material. Fibrin glue is popular because it is completely biodegradable and can be injected before it becomes solid. Unfortunately, the mechanical strength of fibrin glue is weak, so its use as a primary scaffold in articular cartilage engineering is limited. Because of this, fibrin is often combined with other materials to help retain its structure. Chondrocytes have been seeded in pure fibrin glue [332], as well as in mixtures with alginate [333, 334] or collagen [335]. Biochemical results did not show major differences from other scaffold materials. However, genipin cross-linked fibrin scaffolds showed accumulation of collagen type II and aggrecan with a corresponding increase in compressive and shear moduli [336].

Hyaluronan (HA) or hyaluronic acid is a polysaccharide that has been used to create biocompatible scaffolds for cartilage engineering applications. HA is a non-sulfated glycosaminoglycan that helps in lubrication of the joint. It can be cross-linked to form a scaffold capable of supporting chondrocytes. Similar to fibrin glue, HA is injectable and performs well as a minimally invasive approach to filling irregularly shaped defects. However, hyaluronan has also been investigated for use as a solid, porous scaffold. Scaffolds made of an HA-derivative that were implanted *in vivo* showed good histological results for cartilage matrix deposition [337]. Other researchers have found that cross-linked HA sponges produced better histological results than benzylated HA, which was, in turn, better than untreated defects [338]. Integration with the host tissue improved in conjunction with histological findings.

A current hypothesis in the field of tissue engineering is to use scaffold materials made from the same molecules as that of the tissue being repaired [339]. One method to achieve this is through reconstituted matrices. Early instances of using homogenized tissue samples to stimulate regeneration are found with bone therapy and bone tissue engineering [340]. More recent examples include decellularized heart and tracheal tissues [341, 342]. Growth factors and other bioactive molecules resident in harvested tissue are hypothesized to promote the formation of new matrix. Reconstituted matrices use this material to form scaffolds that allow for cell seeding and tissue growth. For example, cartilage tissue can be harvested, homogenized, washed, and then frozen and lyophilized to create sponge-like scaffolds that promote chondrogenesis in adult stem cells [343]. Protein mixtures are another variation on using naturally secreted matrix molecules as scaffold materials. Matrigel, a commercially available product, has been used extensively as a model basement membrane for biological experiments [344]. Scaffolds formed from materials already present in the body present an attractive approach to facilitating the natural regeneration process without eliciting an immune response.

3.3.2   SYNTHETIC SCAFFOLDS

Synthetic scaffold materials are fabricated commercially or in a laboratory, and unlike natural polymers, can be customized in terms of their physical and chemical properties. Specific characteristics of a polymer, such as its mechanical strength and degradation profile, can be altered through modification of its chemical composition. This flexibility allows researchers to design scaffolds with known degradation rates, biological activity, or specific mechanical characteristics.

Poly-glycolides, poly-lactides, and their copolymers are commonly used for scaffold materials and other biomedical applications [345–350]. These polymers can be formed into porous scaffolds, non-woven meshes, or felts, which allow numerous possibilities for scaffold shape and architecture. Poly-glycolic acid (PGA), perhaps the most commonly used synthetic polymer in cartilage engineering, is an alpha polyester that degrades by hydrolytic scission. Total degradation can occur within four to twelve months, which is brief compared to other implanted polyesters [351]. Loss of mechanical properties occurs prior to this, sometimes as early as a few weeks. Since the degradation products of PGA are naturally resorbed into the body, it is attractive for many medical applications requiring biocompatibility. PGA can be formed into a porous scaffold by applying a salt-leaching process. The porosity and interconnectivity of the pores can be controlled by adjusting the amount of salt included during fabrication. PGA is often extruded as thin strands ($\sim$13 μm in diameter) that can be used for making sutures and threads or weaving three-dimensional structures [352]. For cartilage engineering purposes, however, PGA is more commonly used in non-woven mesh or felt forms. The porosity in mesh scaffolds is high, allowing good nutrient transfer throughout the construct. Furthermore, the interconnectivity of the pores increases seeding efficiency since cells can infiltrate throughout the scaffold. One major drawback to these mesh scaffolds is their mechanical functionality. The initial scaffold structure is too weak to be immediately used in loading-bearing environments. However, growth of neocartilage in the scaffold pores is hypothesized to compensate for the mechanical deficiencies of the scaffold itself. Over time, secreted matrix should fill the void space, giving the construct sufficient mechanical integrity to withstand the joint environment. Consistently good extracellular matrix production has been observed using PGA scaffolds, which, along with predictable degradation rates, makes PGA attractive for cartilage engineering experiments [352–354]. Past studies have also shown that PGA promotes more proteoglycan synthesis than other materials like collagen or poly-glycolide/lactide copolymers [355]. While the predictable degradation profile of PGA is often seen as a positive trait, it can also cause problems for tissue engineering applications that require scaffold integrity longer than a few months. For these applications, other polymers have been investigated.

Another alpha polyester polymer used extensively in the medical field is poly-lactic acid (PLA), which like PGA, has been approved by the FDA for implantation in humans. PLA generally degrades slower than PGA with a total degradation time ranging from twelve months to over two years [356]. Again, the loss of mechanical properties and scaffold integrity occurs prior to this, which could cause an engineered construct to fail prematurely. As with PGA, the degradation products of PLA are resorbable, making it an attractive, biocompatible material for implantation. PLA exists in two stereoisometric forms, giving rise to four different types of PLA: poly-D-lactide, poly-L-lactide, poly-D,L-lactide, and poly-meso-lactide [357]. Applications of the PLA isomers range from drug delivery to suture materials. However, the D and L monomers polymerize to form semicrystalline structures that have been investigated as possible scaffolds for cartilage engineering. PLA scaffolds are primarily made as non-woven meshes due to the previous success of this structure for neocartilage formation. Studies have shown that chondrocytes might not have as great of an affinity for PLA surfaces as PGA surfaces, but, over time, total cell numbers on the two materials are similar [358]. Due to its slower degradation rate, PLA scaffolds allow more time for matrix formation before catastrophic loss of mechanical integrity. This is important for applications where the scaffold has to bear loads for a significant period after implantation.

Poly-lactic-co-glycolic acid (PLGA) is a copolymer composed of PGA and PLA monomers. The material properties of PLGA are dependent on the ratio of each monomer included in the macromolecule. For example, a formulation with a large fraction of PLA will degrade slower than one with a large fraction of PGA. Characterization of a 75/25 (PLA/PGA) copolymer of showed a degradation time of 4-5 months, whereas a 50/50 copolymer degraded in only 1-2 months [356]. As with its base components, PLGA degrades into molecules that naturally resorb in the body. General biocompatibility has been investigated in large and small animal models, as well as in clinical trials [346,359]. PLGA has been used extensively as a suture material due to its high tensile strength and controllable degradation rates. It can be fabricated in forms similar to PGA and PLA [360], with the non-woven mesh being among the more preferred structures for recent cartilage engineering studies.

Another popular synthetic polymer for cartilage engineering is poly-caprolactone (PCL). This polymer possesses longer degradation times than PGA/PLA/PLGA and is generally stronger, making it attractive for many orthopaedic applications [351]. PCL can be extruded into threads for meshes/felts or formed into porous scaffolds through a salt-leaching process. As with other polyesters, PCL degrades through hydrolytic scission, but this process can take one or two years to completely degrade the material [356]. While this means the scaffold remains at the implant site, it can also help with the mechanical integrity of the construct during early time periods. The resistance of PCL to rapid hydrolysis is an attractive trait for some applications. Copolymers including PCL incorporate the strength and elasticity of the material while allowing slightly faster degradation times [358]. Poly-L-lactide-epsilon-caprolactone implanted into mice showed formation of cartilage-like structures after four weeks, with minimal degradation [361].

A more recent trend for synthetic polymers is to fabricate materials that can control the attachment of cells and proteins to the scaffold. The most common method to accomplish this is to modify the hydrophilicity/hydrophobicity of the material. Poly-ethylene glycol (PEG) is a polymer that prevents adsorption of proteins and cells due to its high hydrophilicity. PEG can be incorporated into copolymers and thereby modify the cell attachment characteristics of a material. This property can be used to allow cell attachment on only certain portions of an implant or no cell attachment at all. The latter is one reason PEG is often used in copolymers – to improve biocompatibility [362]. The incorporation of PEG molecules increases hydrophilicity, which helps to prevent adsorption of antibodies and other proteins, thereby lessening any immune response. PEG by itself has similar mechanical properties in compression to cartilage, with higher modulus values corresponding to higher molecular weights [363]. It has been copolymerized with a number of different materials to take advantage of its biocompatibility traits to create materials for a variety of applications [317, 362, 364–370]. Copolymerization is also necessary since PEG does not naturally degrade in the body, an attribute necessary for long-term success of an implanted construct. For articular cartilage engineering, degradation of the scaffold is necessary to provide space for new tissue to form.

An alternative approach to synthetic polymers is the creation of macromolecules that imitate natural biomaterials. Researchers have successfully synthesized genetically engineered molecules, such as elastin-like polypeptide (ELP), that are similar to natural proteins found in the body [371]. Chondrocytes cultured in the gelled form of ELP maintained their phenotype, secreting matrix molecules such as sulfated glycosaminoglycans and collagen.

3.3.3   COMPOSITE SCAFFOLDS

This could include a fiber scaffold formed from several different natural and synthetic threads or a naturally derived hydrogel infused throughout a synthetic mesh. Composite scaffolds consist of two or more of the previously discussed materials incorporated into a single scaffold. This could include a fiber scaffold formed from several different natural and synthetic threads or a naturally derived hydrogel infused throughout a synthetic mesh. For example, the void fraction of PLGA meshes can be filled with chondrocytes encapsulated in fibrin glue, which allows for a rounded cell phenotype, good cell distribution throughout the scaffold, as well as tunable degradation characteristics. This approach produced 2.6 times more GAG after 4 weeks than PLGA alone [372]. The inclusion of fibrin glue might have helped retain GAG molecules in the construct, whereas GAG simply diffused out of the bare PLGA scaffolds.

Infiltrating a fiber-based scaffold with a hydrogel is a popular form of composite scaffold. However, the cell-material interactions are critically important to the overall success of the construct. In previous studies, chondrocytes encapsulated in alginate were combined with either PLGA or demineralized bone matrix (DBM) before implantation into mice for eight weeks [373]. The PLGA-alginate composite produced collagen type II, a positive indicator of cartilage formation. The DBM-alginate composite, however, did not produce collagen type II. The cell response could be modified by substituting other hydrogels or including growth factors, but the base scaffold materials will still play a major role in the type of matrix deposited in the construct.

Another approach to composite materials is to reinforce solid scaffolds with fibers oriented in specific directions. By embedding fibers in a scaffold, the mechanical properties can be modified to improve strength in preferred directions. This is particularly important for anisotropic tissues such as articular cartilage. Fiber reinforced scaffolds can be fabricated using any combination of materials. Past studies have investigated PGA fiber-reinforced PLGA and found that the compressive modulus and yield strength improved by up to 20% [374]. Carbon fibers, while seemingly unadvisable for joint implantation, have been used with satisfactory clinical results for filling defects *in vivo* [375, 376]. Success rates of 70-80% were achieved based on qualitative measures of pain several years after implantation.

Composite scaffolds that incorporate several different types of materials can help replicate the complex structure necessary for providing functional properties appropriate to load-bearing tissues [377]. One approach that shows promise is to fabricate three-dimensional structures that exhibit mechanical properties similar to articular cartilage immediately after implantation. Woven scaffolds, such as alginate-filled PCL meshes [292], can provide both mechanical strength, anisotropy, and a beneficial growth environment.

3.3.4   SCAFFOLDLESS

Though scaffolds can serve as an additional tool in controlling tissue development (e.g., with the slow release of growth factors or pre-patterned to influence organization), they also bring with them issues such as degradation toxicity, stress shielding, and cell signaling hindrance. Techniques have also been developed using chondrocytes to create scaffoldless constructs [312, 378–381]. Via centrifugation, chondrocyte pellets can be formed and grown in culture. With gentle or no fluid movement (e.g., rotational culture [382] or low density seeding on agarose [383]), larger aggregates have also been formed. It was proposed that the formation of numerous aggregates may serve as a 3-D culture methodology for chondrocyte expansion [382] while aggregates of limb bud cells have been used to examine parallels in development [383].

Emerging from the developmental studies, scaffoldless culture has been proposed as a method to engineer functional articular cartilage of sufficient dimensions. For instance, a self-assembly method has been developed based on the differential adhesion hypothesis to produce robust cartilage constructs that contained two thirds more GAG than native tissue, and collagen levels that reached one third the amount of native tissue. Neocartilage thus formed contain collagen type II and chondrocytes in lacunae (Figure 3.1). More importantly, the compressive stiffness of self-assembled cartilage reached more than one third of native tissue values [313].

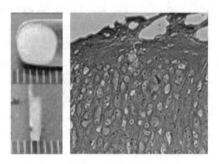

Figure 3.1: Gross appearance of 6 mm self-assembled discs punched from 15 mm constructs (left panel). Each mark is 1 mm. Constructs stain throughout for collagen type II, with zonal morphological appearance, and chondrocytes residing in lacunae (right panel).

Within a scaffoldless system, increased N-cadherin expression during neotissue formation suggested that differential adhesion mediated self-assembly [314] while another study has shown that chondrocytes assemble via β1 integrins [384]. The mechanical properties of scaffoldless cartilage has been studied [385, 386], and, within scaffoldless cartilage, several biochemical properties have recapitulated cartilage development. Increased proportion of collagen type II, decreased proportion of collagen type VI, decreased chondroitin 6- to 4- sulfate ratio, and localization of collagen type VI to the pericellular matrix (Figure 3.2) [314] are all evidence of maturation in scaffoldless cartilage constructs. These studies showed that the self-assembly method mimicked tissue development and maturation, suggesting that a set of exogenous stimuli could then be applied to augment tissue functional properties.

Various growth factors have been applied individually and in combination to self-assembled constructs, with TGF-β1 showing the greatest potency, inducing 1-fold increases in both aggregate modulus and tensile modulus, and increasing GAG and collagen content [387]. Scaffoldless constructs have also been cultured under various mechanical forces. Hydrostatic pressure stimulation [388], shear and compression [389], and other forms of bioreactor culture [382] have shown to be advantageous for scaffoldless constructs. Combined regimens of different classes of stimuli (biochemical and mechanical) have also been examined to show additive and synergistic effects on the functional properties of scaffoldless cartilage [390]. Scaffoldless cartilage has also been implanted in goats [391] to demonstrate articular cartilage resurfacing when used in combination with a periosteal flap.

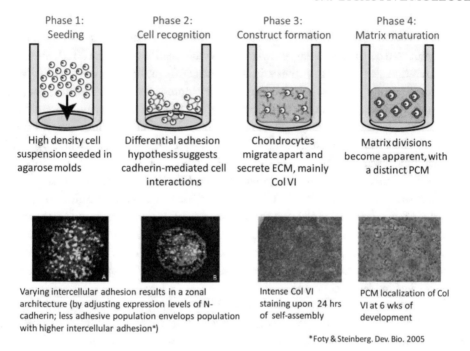

Figure 3.2: Milestones in the self-assembly of differentiated articular chondrocytes mimic those seen during cartilage development. (Used under the Creative Commons Attribution License, from Ofek et al., PLoS ONE, 2008 [314].)

3.4 BIOACTIVE MOLECULES

Growth and development of cartilage tissue relies heavily on biochemical signals. The sequence, duration, and intensity of stimulation can all play roles in how cells secrete matrix in a regenerating environment. Bioactive molecules can include growth factors, adhesion proteins, peptide sequences, or any other entity that binds to cells to create a biological response. There are more bioactive molecules present in the body than this chapter can easily encompass, so only a few of the more common growth factors, proteins, and peptides will be highlighted. All have shown proven effects for cartilage engineering, and future work will certainly make use of them to accelerate regeneration of functional tissues. Other less often examined factors can encompass those described in Section 2.1, which discusses cartilage formation during development. The following chapter will focus initially on growth factors, specifically the transforming growth factor-β (TGF-β) superfamily, followed by a description of possible scaffold modifications using bioactive molecules.

3.4.1   GROWTH FACTORS

The inclusion of stimulatory growth factors is one of the most common means to accelerate tissue growth in engineered constructs. Many growth factors have been shown to be effective at stimulating cellular proliferation and matrix synthesis in articular cartilage, both *in vitro* and *in vivo*. Since growth factors normally play a role in healing and development, their therapeutic use is intended to replicate this function to promote rapid regeneration of a tissue. Varying amounts of growth factors are constantly present throughout the body, so higher concentrations are typically used in experiments to elicit more dramatic effects. For example, a simple media cocktail with 20% fetal bovine serum added was shown to have similar effects on proliferation and protein synthesis as a cocktail that included three different growth factors (TGF-β1, basic fibroblast growth factor (bFGF), and insulin-like growth factor-1 (IGF-1)) [392]. Lower concentrations of serum in the media produced poorer results, indicating the importance of growth factors that circulate naturally in the body. The problem with serum, however, is that its composition and the concentrations of its components are generally unknown and can vary widely from source to source and batch to batch. Growth factors are a means to stimulate a response using controlled amounts from an external source, which is important for safety concerns.

Growth factors can have a synergistic relationship with mechanical loading, which is a boon for tissues in load-bearing environments such as articular cartilage. *In vitro* experiments investigating the effect of both growth factors and mechanical stimulation have shown significant increases in matrix production compared to either stimulus alone [199, 393, 394]. Notably, such synergism has been shown to affect functional properties in scaffoldless constructs formed by the self-assembly process. A 164% increase in the aggregate modulus value (H_A), a 231% increase in the Young's modulus value (E_Y), an 85% increase in glycosaminoglycan/wet weight, and a 173% increase in collagen/wet weight relative to controls can be achieved when combining growth factors with mechanical stimulus (in this case, TGF-β1 and hydrostatic pressure) [390]. This relationship extends to the *in vivo* environment as well. Implanted, growth factor-laden collagen sponges showed either bone-, cartilage-, or tendon-like tissue formation depending on the loading environment [395]. Many studies have also been performed with growth factors alone, without interaction with mechanical stimuli. Dramatic effects on proliferation, differentiation, and synthesis have been documented, for example, for IGF-1, bFGF, hepatocyte growth factor (HGF), and platelet-derived growth factor (PDGF) [396]. Of particular interest is the TGF-β superfamily, which includes bone morphogenetic proteins (BMPs). This group of bioactive molecules has been shown to significantly affect chondrogenesis and bone growth, both of which are important for successful regeneration of osteochondral defects [397].

The TGF-β superfamily is a class of growth factors that is involved in the repair and inflammation response following injury [398]. Numerous studies have shown that these growth factors can also elicit dramatic changes in articular chondrocytes. TGF-β1 is a popular isoform used in cartilage engineering studies with some indicating stimulation of chondrogenesis and proliferation [394, 399, 400] and others showing inhibition of matrix formation [401, 402]. These mixed effects could be caused by various experimental factors in the different studies as the biological response of cell populations can be quite complicated. TGF-β1 has been shown to promote collagen formation and increase construct wet weight after four weeks *in vitro* [394, 399]. However, the effectiveness of the growth factor could be dependent on the differentiation state of the cells. For example, TGF-β1 stimulated proliferation and proteoglycan synthesis in chondrocytes that have been cultured for a week *in vitro*, but these effects were not apparent for freshly isolated chondrocytes [401]. Additionally, arthritic chondrocytes experienced a decrease in proteoglycan synthesis when treated with TGF-β1 [402]. *In vitro* culture and disease have both been shown to affect the normal phenotypic expressions of chondrocytes, so their response to biochemical stimuli is not surprisingly altered in comparison to healthy cells *in vivo*.

BMPs play a major role in endochondral bone formation and show general effects on cellular proliferation and matrix synthesis. As explained above, they are particularly attractive for cartilage engineering studies because they affect both chondrogenesis and osteogenesis. Osteochondral integration is a critical factor in whether implants succeed or fail *in vivo*, so molecules that can stimulate this response are desirable [403]. As with TGF-β, BMPs can act synergistically with mechanical stimuli to accelerate regeneration of joint tissues. Currently, 20 types of BMPs have been identified, but only a subset has been examined for cartilage regeneration. BMPs generally have the ability to guide stem cells and immature bone and cartilage cells along the osteochondral pathway [404]. In experimental studies, BMP-1 showed greater stimulation of proteoglycan and collagen synthesis than TGF-β1 [404]. BMP-2 upregulated proteoglycan and collagen expression in chondrocytes [403, 405, 406] while also inducing better healing of defects *in vivo* [398, 403]. BMP-4 showed an ability to stimulate proteoglycan synthesis, bone formation, and cellular proliferation [403, 407]. BMP-7 also showed positive effects on matrix synthesis [403] and proliferation [408] while also decreasing collagen type I expression and suppressing infiltration of fibroblasts *in vivo* [409]. Articular chondrocytes treated with either BMP-12 or -13 synthesized elevated levels of GAG although these increases were less than that observed for cells treated with BMP-2 [405]. Overall, experimental results have shown that BMPs have a generally positive effect on cartilage differentiation and morphogenesis whether alone or in combination with other growth factors. For example, BMP-2 application with IGF-I resulted in over a 1-fold increase in aggregate modulus, accompanied by increases in GAG production, as compared to controls [387].

The TGF-β superfamily also includes several other groups of growth factors known to affect cartilage growth and differentiation. Cartilage derived morphogenetic proteins (CDMPs), osteogenic proteins (OPs), and growth differentiation factors (GDFs) have all been investigated as possible means to accelerate regeneration of joint tissues *in vitro* and *in vivo*. Some of the growth factors included in these groups are actually the same molecule. For example, the pairs OP-1/BMP-7, CDMP-1/GDF-5, and CDMP-2/GDF-6 are the same growth factors with alternate designations. All of these molecules can affect chondrocytes in a manner similar to other TGF-β superfamily members. OP-1, CDMP-1, and CDMP-2 all increase proteoglycan synthesis and cellular proliferation [410] although OP-1 was found to be the more effective stimulus [411]. GDF-5, which is naturally present in articular cartilage, also increases proteoglycan synthesis [412].

Growth factors are typically delivered as soluble components in culture media cocktails. While this is acceptable for *in vitro* experiments, delivery becomes more complicated once the construct is implanted. An alternative approach is to include polymeric carriers, such as microspheres, in the construct so that growth factors are released over time [334]. Early approaches found little success since most polymer release times lasted only a few days. Once freed, growth factors can degrade within a week, so long-term treatments using these carriers would be infeasible [413]. However, current research has indicated that alternative polymers, such as elastin-like polypeptides, have the capability to extend the release time of drugs to weeks or months [414]. These carriers would allow long-term stimulation of the implant with a local source of growth factors, further stimulating matrix growth and possibly helping integration with the surrounding tissue.

Another alternative to long-term growth factor stimulation is to modify the gene expression of implanted cells using either transfection or other forms of genetic modification [415, 416]. In this case, growth factors are secreted by cells within the defect site. Local stimulation means that implant growth and integration do not rely on externally provided drugs. Many different approaches can be used to apply this technique. For example, all implanted cells could be modified or only a fraction. Alternatively, the modified gene could be conditionally active, which would be advantageous if stimulation is only desired for certain periods during regeneration. While this approach seems attractive, the practicality of genetic modification creates a large barrier to its current success. Total control of how the gene is expressed is not currently feasible, which creates a safety issue. Furthermore, the long-term effects of elevated growth factor levels are not known, especially on neighboring tissues that are not involved with the cartilage repair process.

Growth factors do not necessarily need to be available as unbound molecules to induce a response from resident cells. Modifying the scaffold material itself with growth factors is a possible means to stimulate cells growing in the construct. Proteins such as TGF-β, IGF, PDGF, HGF, and FGF have all been used in such modifications. The benefit of scaffold-bound growth factors for *in vivo* experiments is clear, but even *in vitro* experiments could be enhanced since bioactive molecules would be distributed evenly throughout the scaffold, affecting all regions equally. Without restriction of the growth factors, diffusion could result in loss of bioactive molecules to the surrounding environment.

There are two primary methods for including growth factors in engineered constructs. The first, encapsulation, was mentioned previously and involves sequestering stimulatory proteins in polymeric materials that have controllable release characteristics. Alternatively, a growth factor can be encapsulated in the bulk scaffold material, assuming that the scaffold has characteristics that can restrict the diffusion of certain molecules out of the construct. For example, hydrogels can restrict the diffusion of growth factors for short periods of time in comparison to meshes or felts. Incorporating multiple types of materials in a scaffold is another means to allow drug release over time. A two-phase PLGA implant loaded with TGF-β showed good results when implanted into osteochondral defects [346]. Microparticles are the most common methodology for including growth factors in scaffolds with high porosity and large pore sizes such as meshes and felts [417]. Since the degradation characteristics of the carrier polymer can be customized, protein concentrations are more predictable.

The second method for incorporating growth factors is to bond the bioactive proteins to the surface of the scaffold material. Covalent bonding is one approach that has been used for cartilage engineering. The major drawback to this methodology is that immobilizing proteins often decreases their effectiveness. The active regions of a molecule can be obstructed once bound to a surface although this is dependent on the protein being bound and the chemical reaction used to form the covalent bond. Past studies have proven it is possible, though, with one group showing significant stimulation of cells cultured in a PEG hydrogel that included covalently bound TGF-β1 molecules [418]. Tethering of growth factors to the scaffold material is a promising means of retaining activity of the protein while still restricting its movement within a scaffold [419]. With this approach, the growth factor is attached to a molecular chain that extends away from the surface, allowing greater access while still restricting its diffusion in the construct.

The dramatic stimulation provided by growth factors in cartilage engineering studies suggests that their inclusion may be required for successful regeneration of the tissue. However, the most effective means to apply them is still to be determined. Furthermore, the wide variety of growth factors available creates myriad combinations that could accelerate the growth process, but this must be tempered by knowledge of what additional effects each growth factor has on the construct and the surrounding tissues. Cocktails of multiple growth factors might be the best approach to creating a functional tissue. For example, IGF-1 has been used to increase GAG synthesis, TGF-β1 for improving collagen content, and interleukin-4 for minimizing GAG-depleted regions in an engineered construct [394]. Whether growth factors are applied as a soluble mediator, encapsulated in polymeric carriers, or chemically bound to a scaffold surface, they are an integral part of the cartilage engineering process and will continue to be a major area of focus for tissue regeneration therapies in the future.

3.4.2   PROTEIN COATING AND PEPTIDE INCLUSION

Beside attaching growth factors, scaffold materials can be modified via protein coating, peptide incorporation, or micropatterning to alter cell attachment characteristics. The first two of these approaches capitalize on integrin-receptor relationships between cells and extracellular matrix proteins to direct cell attachment in a controlled manner. Chondrocytes express specific integrins that will bind with corresponding proteins, so by coating any material with the correct protein, cells can be made to adhere to almost any bulk structure [420]. Integrins identified on chondrocytes include $\alpha_1\beta_1$, $\alpha_2\beta_1$, $\alpha_5\beta_1$, $\alpha_V\beta_5$, $\alpha_V\beta_3$, and $\alpha_3\beta_1$, with the latter two being more prevalent on superficial zone chondrocytes than deep zone chondrocytes [421, 422]. Biomaterials can be modified with proteins that express one or more of these integrins to control cell attachment to different regions of the scaffold. While binding between cells and proteins is not permanent, it can at least provide anchorage points for cells within the construct.

In addition to cell adhesion, extracellular matrix proteins can also promote the haptotactic and chemotactic motility of chondrocytes. By modifying the adhesion characteristics of a biomaterial, cellular migration into a scaffold can be increased or decreased [422]. Simple adsorption of fibronectin onto a polymer scaffold showed an increase in cell attachment and ingrowth compared to uncoated controls [423]. The benefits of increased cell migration have also been observed for *in vivo* experiments. Hyaluronan scaffolds coated with fibronectin showed increased tissue ingrowth after implantation into osteochondral defects [424]. This ingrowth helped improve integration with the surrounding bone and cartilage, which is very important for the long-term success of an implant.

One of the more common methods to modify the attachment characteristics of a surface is to coat it with proteins. The hydrophobicity or hydrophilicity of a material determines the degree of interaction. Hydrophobic materials, in particular, allow proteins to readily adsorb to their surfaces, with more hydrophobic materials forming a stronger interaction than less hydrophobic materials [425]. Following adsorption, cells can then bind to those proteins that coat the bulk scaffold material. Altering the types of proteins adsorbed to a surface can affect which cells can attach. More generally, controlling whether proteins can adsorb helps with the biocompatibility of an implant. Some of the more common adhesion proteins present in the body include collagen, thrombospondin, osteopontin, bone sialoprotein, fibronectin, vitronectin, fibrinogen, von Willebrand factor, laminin, entactin, and tenascin [420, 426]. Based on the integrin receptors present on chondrocytes, all of the before mentioned proteins promote adhesion except osteopontin, entactin, and tenascin [427]. Typically, collagen and fibronectin are used for cartilage applications although a wide variety of cell types, not just chondrocytes, have shown an affinity for these ubiquitous proteins. In native tissues, extracellular matrix molecules help transmit mechanical and chemical stimuli to cells. Replicating this function in an engineered construct is one of the hopes associated with protein coating. Naturally secreted proteins will likely play a more dominant role as the engineered construct develops, but the initial stages could be assisted by the inclusion of supplementary proteins.

Collagen has been investigated extensively as a bulk scaffold material for tissue engineering and less so as a protein coating. However, monolayer studies have shown that chondrocytes readily attach to collagen surfaces [428], making it a logical choice as a coating on materials that otherwise prevent cell attachment. The type of collagen used could play an important role since the primary collagen in articular cartilage is type II, not type I. Past studies have shown that chondrocytes have a preference for collagen type II fragments over a type I collagen-coated surface, possibly due to the integrin receptors expressed by articular chondrocytes [429]. While beneficial to cell adhesion, collagen alone does not appear to retain chondrocyte gene expression when cells are cultured in monolayer [97]. In general, collagen has shown good results as a matrix material, so its use as a coating has been limited.

Vitronectin is another protein that has been investigated as a scaffold coating for tissue engineering. Past results showed that vitronectin controls osteoblast attachment and spreading, as opposed to a more ubiquitous protein like fibronectin, when used as a coating *in vitro* [430]. This finding was determined by using culture media without fibronectin or without vitronectin. Samples in the prior group showed no effect on cell attachment while the latter group had markedly reduced attachment and spreading. While these findings are not directly applicable to chondrocytes, vitronectin could still be important in modulating the attachment of cells in osteochondral constructs. Other experimental studies investigating vitronectin have shown that it competes better than most proteins when adsorbing to surfaces in the presence of serum [431]. Additionally, good adhesion between vitronectin and chondrocytes has been observed, possibly through the $\alpha_5\beta_1$, $\alpha_V\beta_5$, and $\alpha_V\beta_3$ integrins.

Cartilage matrix protein (CMP), not to be confused with cartilage oligomeric matrix protein (COMP), is a less extensively studied molecule that is expressed almost exclusively in cartilage [432]. CMP binds to aggrecan and collagen type II, and chondrocytes attach to it via the $\alpha_1\beta_1$ integrin during adhesion. When used as a coating material, CMP enhanced both cell attachment and spreading on surfaces [433]. The addition of collagen type II to the CMP coating showed even more improvement in these characteristics. Because CMP is specific to cartilage tissue, it might be a more appropriate protein to target for coating purposes although issues such as ease of production and cost would certainly be important factors.

Functional cartilage constructs will have to be designed as three-dimensional structures, but preliminary experiments can still be conducted in monolayer to investigate areas such as cell-surface interactions. Micro- and nano-technologies now allow for precise control on protein placement on a variety of surfaces. Technologies such as soft lithography and self-assembled monolayers allow protein stamping on materials that restrict the attachment of cells to specific regions [434]. Custom designs incorporating multiple types of proteins are feasible using these techniques, making possible a wide variety of experiments at the cellular level. Many researchers are investigating cell-surface interactions using these patterning techniques, which are hoped to be translatable to three-dimensions in the future.

A major difficulty with using proteins to facilitate adhesion and migration of cells within scaffolds is that the protein-surface interaction is typically transient. Unless the protein is cross-linked to the material, a state that can dramatically affect its biological activity, proteins will eventually disassociate from the scaffold. For short-term applications such as cell seeding, this is not a problem. For longer-term needs, the base scaffold material should be hydrophobic, which increases the affinity proteins have for the surface. While two-dimensional surfaces can be rigidly designed to control where cells attach, three-dimensional scaffolds are more difficult. Choosing an appropriate base material is the best starting point if the experiment relies on protein coating over longer time frames.

An attractive alternative to protein coating is to use only the amino acid sequences, or peptides, involved in cell-surface binding. This approach allows for permanent modification of a scaffold since the peptides are chemically bonded to the material rather than just adsorbed. The number of binding sites and their location can be controlled during fabrication, and these parameters do not change, provided the scaffold does not degrade or otherwise alter its structure dramatically. Cells attach to protein-coated surfaces through amino acid recognition sequences located in the macromolecular structure. Peptides are simply short amino acid sequences derived from the larger protein. After modification, materials that are otherwise unattractive to cells can now be successfully seeded and used in tissue engineering applications. Peptide sequences are also proposed as a means to modify cellular gene expression and protein synthesis, not just for controlling cell attachment. For example, if only the functional sequence associated with a growth factor binds to a cell, then the resulting response is expected to be similar to binding the entire molecule since the same pathways are activated in both cases.

Peptides are typically grafted to a material by covalent bonding, which securely attaches the sequence to a location and prevents diffusion through the construct or disassociation from the surface. However, as noted previously, chemical bonding can have the negative effect of reducing the biological activity of attached peptides. Tethering the molecules via a linker chain can help prevent this by moving the peptide away from the surface, allowing more flexibility in its binding configuration with cells [420]. Because some base scaffold materials can sterically hinder biological reactions, the linker chain must be long enough to allow cell integrin-peptide binding away from the surface. Typically, peptides are grafted to biomaterials that otherwise prevent the attachment of cells and proteins. Therefore, scaffolds can be designed that exhibit attachment characteristics entirely dependent on the types, concentration, and location of peptides bound to their surface.

The peptide density on a material controls not only cell attachment, but also cell motility. While increasing the concentration of peptides can increase cellular attachment, it can also decrease the ability of cells to migrate since they are adhered to the surface at numerous binding sites [420]. Balancing these two parameters is difficult and is highly dependent on cell type and application. If migration of cells into a scaffold is desired, then high densities of peptides cannot be used. While this could reduce cell attachment during seeding, other parameters such as the duration of seeding or total cell numbers used might compensate for the lower seeding efficiency. Alternatively, integrin clustering can be used to help facilitate migration without totally restricting cell motility [422]. This approach is especially important for applications that require ingrowth of surrounding cell populations into the construct.

The most common peptide sequence used for cell attachment is Arg-Gly-Asp (RGD), which was originally identified as a recognition sequence located on the larger fibronectin molecule [420]. Further investigation has shown that it is a ubiquitous peptide found in many species of both plants and animals, indicating its importance and longevity in evolution [435]. RGD exhibits an ability to bind with 8-12 integrins (out of 20+ currently identified), making it incredibly useful for tissue engineering applications [426]. In chondrocytes, RGD binds strongly to $\alpha_5\beta_1$, $\alpha_V\beta_5$, and $\alpha_V\beta_3$ and weakly to $\alpha_3\beta_1$, making it an attractive sequence for cartilage engineering studies.

Many different peptides have been investigated for their use in tissue engineering and drug delivery applications. Peptide binding is associated with cellular integrins, so some sequences are only useful for specific cell types. Besides RGD, other peptides that have known activities are: KGD ($\alpha_{IIb}\beta_3$), PECAM ($\alpha_V\beta_3$), KQAGDV ($\alpha_{IIb}\beta_3$), LDV ($\alpha_4\beta_1$ and $\alpha_4\beta_7$), YGYYGDALR and FYFDLR ($\alpha_2\beta_1$), and RLD/KRLDGS ($\alpha_V\beta_3$ and $\alpha_M\beta_2$) [426,435,436]. Other sequences have also shown functionality although researchers are still investigating the binding relationships involved in each case. Chondrocytes have shown a strong binding affinity for RGD, PECAM, YGYYGDALR, FYFDLR, and RLD/KRLDGS with weak binding for KQAGDV and LDV [426].

Functional peptides are not restricted to a set number of amino acids. Long sequences corresponding to specific attachment proteins have been used successfully to modify scaffold materials. However, the key sequence (e.g., RGD) must be accessible to cellular integrins to allow binding. Sequences such as **GRGD** [437], **GRGD**SP [438], and CGGNGEP**RGD**TYRAY [439] all use the RGD sequence to facilitate cell binding, but the peptides are modeled on different proteins (the latter is from bone sialoprotein). Shorter peptide sequences are often preferred because of their versatility. For example, GRGD can be synthesized onto the end of hydrophilic linker chains that are attached to an underlying bulk material, thereby allowing cell seeding on scaffolds that are otherwise unattractive [437].

While most studies involving peptides have been conducted in monolayer, their use is not restricted to two-dimensions. A number of researchers have been investigating how peptides can be incorporated into three-dimensional scaffolds for use in tissue engineering [440–442]. Peptides can be incorporated into hydrogel scaffolds or grafted onto the exposed surfaces of porous scaffolds. As in monolayer, cells are expected to bind to available peptides, thereby altering cellular proliferation, migration, and differentiation. Alginate modified with RGD has been shown to promote cell adhesion, spreading, and chondrocytic differentiation [443]. However, alternative studies using adhesion peptides and PEG hydrogels showed a reduction in proliferation and protein synthesis [444, 445]. These discrepancies could be caused by differences in peptide densities, which are more difficult to control in three-dimensions than in two-dimensions.

As with full proteins, micropatterning techniques can be applied using peptide sequences to create specific designs on material surfaces [446–449]. These experiments typically focus on the attachment properties of cells since only a small portion of the protein structure is actually present. Integrin binding reactions can be investigated in a controlled environment using this experimental setup. Additionally, micropatterning can be used to control the geometry of single cells, allowing investigations of cytoskeletal structures. Interactions between different cell populations have also been studied using monolayer micropatterning [450]. Similar to protein stamping, peptide-patterned regions allow cell attachment whereas the rest of the surface does not. Theoretically, if different peptides are patterned in specific regions, then only cells expressing matching integrins will be able to bind, creating a surface with segregated populations based on cell phenotype. This approach would be interesting for co-culture experiments, assuming the populations are different enough to possess distinct integrin profiles.

Micropatterning can control cell morphology by restricting the available surface binding sites. Cell shape has been shown to influence whether a cell will proliferate, die, or differentiate [449]. For some cell types, a spread/flat morphology promotes proliferation while a severely restricted, rounded morphology promotes apoptosis and cell death. However, patterned surfaces that fall in between and promote neither growth or death have been shown to induce cell differentiation [448]. This state is different for each cell type and possibly each individual cell since peptide spacing would be largely dependent on cell size. If binding sites are distributed evenly across a surface at an average density, then cells should spread, proliferate, and migrate on the material with ease. If high densities of binding sites are placed in small areas, then a rounded morphology will result instead. Both systems can be used to study different aspects of cell-biomaterial interactions.

Micropatterning is not limited to stamping peptides and proteins. It can also alter the topography of a surface, which in turn, can affect cell attachment, proliferation, and gene/protein expression [451]. By controlling how strongly a cell is bound to the surface, micropatterning affects the migration and proliferation of attached cells. In general, rough surfaces at the sub-micron scale allow for weak cell attachment but inhibit extensive spreading whereas smooth surfaces promote strong attachment and spreading as well as proliferation and migration. Aligned topographies have also been shown to affect cell morphology and differentiation [452].

3.4.3 CATABOLIC AND OTHER STRUCTURE MODIFYING FACTORS

While it is counterintuitive to apply catabolic factors to fabricate a piece of tissue, the enzyme chondroitinase-ABC (C-ABC) has been applied to deplete GAG content to subsequently improve biomechanical properties. C-ABC increased tensile properties of self-assembled articular cartilage without compromising compressive properties as GAG levels return post-treatment [453]. Multiple C-ABC treatments further increased tensile properties, reaching values of 3.4 and 1.4 MPa for the tensile modulus and ultimate tensile strength, respectively [454]. C-ABC represents an exciting method for engineering functional articular cartilage by departing from conventional anabolic approaches. Another structure modifying agent is lysyl oxidase, which acts to crosslink collagen [455]. Attempts to affect collagen crosslinking (and thus mechanical properties) have targeted this enzyme with an inhibitor, beta- aminopropionitrile [456–458].

3.5 MECHANICAL STIMULATION

Bioreactors are a critical component for growing a mechanically functional tissue. For articular cartilage, compressive, tensile, and frictional properties are of the utmost importance, as are the tissue's general wear characteristics, so an *in vitro* tissue engineering approach should focus on improving these properties before implantation. Direct compression and, especially, hydrostatic pressure have been shown to help stimulate the secretion of proteins that are necessary for compressive strength. Low- and high-shear bioreactors also have shown promise in growing functional constructs, perhaps due to the increased nutrient transfer during stimulation. While mentioned here, these bioreactors are discussed in more detail in the next chapter.

3.6 CHAPTER CONCEPTS

- While the *in vivo* environment has been thought to contain all the necessary factors (though not necessarily the cells) in effecting cartilage repair, *in vitro* tissue engineering is gaining popularity due to the well-controlled environment it offers.

- The use of cells, from differentiated chondrocytes to expanded cells and various adult and embryonic stem cells, for *in vitro* cartilage tissue engineering seeks to improve upon or replace the function of injured cartilage tissue.

- Desirable characteristics for cartilage repair scaffolds include biocompatibility, the capacity to bear load, cell attachment, proliferation, and metabolism, and a degradation rate that matches tissue formation.

- Scaffolds can come in several natural or synthetic forms.

- Natural scaffolds that have been studied for cartilage engineering include collagen, alginate, agarose, chitosan, fibrin glue, hyaluronic acid-based materials, and reconstituted tissue matrices.

- Synthetic materials include poly-glycolides, poly-lactides, poly-caprolactone, poly-ethylene glycol, and many others.

- Natural and synthetic materials can be combined to form composite scaffolds, such as infiltrating a fibrous mesh with a hydrogel to form one construct.

- Cartilage tissue engineering has also been attempted without the use of scaffolds in combination with biochemical and biomechanical stimuli.

- Self-assembly, a scaffoldless tissue engineering approach, has recently been demonstrated to result in neocartilage of clinically relevant dimensions and with functional properties approaching those of native tissue.

- Bioactive molecules can be soluble or tethered, with intended effects being anabolic or catabolic, or even structural, in improving the functional properties of engineered cartilage.

- TGF-β, BMP, IGF, bFGF, and many other growth factors in their soluble forms have been applied in different concentrations, dosage frequencies, and combinations to engineer cartilage.

- Proteins and peptide coatings and modifications have been used to improve chondrocyte response to biomaterials. These include collagen, vitronectin, CMP, and many amino acid sequences such as RGD.

- Micropatterning can apply proteins and peptides to effect phenotypical and morphological changes in chondrocytes and other cells.

CHAPTER 4

Bioreactors

Articular cartilage is a mechanically-sensitive tissue that can respond favorably or unfavorably to biomechanical stimuli. The results from experimental studies included in this section combine many different approaches for enhancing cartilage regeneration. However, direct comparisons are difficult due to the variety of cell sources, media cocktails, and general laboratory practices used from study to study. In general, cell-seeded scaffolds, or "constructs," are cultured *in vitro* to help produce a neo-tissue that has sufficient mechanical and/or biochemical properties for implantation.

Although the precise signaling pathways involved in the mediation of mechanostimuli are not completely understood, evidence suggests that certain types of forces are desirable for cartilage synthesis and modeling. Under static conditions, chondrocytes synthesize a material that has poor tissue organization [459]. Since static culture conditions appear to be inadequate, dynamic culture conditions have been studied extensively for their beneficial effects on cartilage synthesis and organization. The creation of cartilaginous material in bioreactors has proven to be successful and is a promising means to obtain reproducible tissue constructs [353, 459]. In general, the composition, morphology, and mechanical properties of cartilage synthesized in a bioreactor appear better than cartilage grown under static conditions [460].

This section includes descriptions of past and current bioreactors used for stimulating cartilage explants or engineered constructs (Figure 4.1). The categories included in this chapter's review of biomechanical stimulation include:

- Direct compression.

- Hydrostatic pressure.

- Shear bioreactors (surface/fluid-surface shearing, direct flow and fluid perfusion, low-shear "microgravity" bioreactors).

- "Low-shear" systems (enhanced nutrient transport, "microgravity" bioreactors).

- Hybrid bioreactors incorporating multiple loading regimes.

4.1 DIRECT COMPRESSION

Compressive loading is a major component of normal mechanical stimulation within diarthrodial joints. The points of contact between the femoral condyle and the tibial plateau (and intervening menisci) cause compression within the cartilage tissue that is separate from other types of mechanical stimulation. Studies focusing on direct compression typically use platens that physically touch the construct surface. Static or dynamic loading with these devices mechanically deforms the sample.

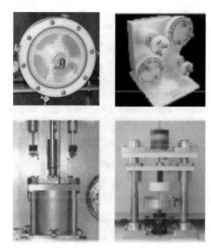

Figure 4.1: Low shear bioreactors (top), hydrostatic pressure (bottom, left), and direct compression (bottom, right) bioreactors.

Direct compression has been used successfully to stimulate the chondrocytic phenotype of cells *in vitro*. The mechanism for this positive response is not fully known. While the actual mechanical stimulation is undoubtedly a major factor, enhanced nutrient transfer and removal of waste products could also contribute to increases in cartilaginous matrix secretion.

Proteoglycan synthesis has often been used in studies as an indicator that a mechanical stimulation is beneficial although some studies also use DNA synthesis, collagen synthesis, or mechanical properties as gauges. Synthesis is not necessarily a measure of composition, though, and indicates only which molecules are made and not whether they are integrated into the tissue. The incorporation of radioisotopes is another measure of macromolecule formation and is an indirect indication of synthesis. Using the reported values of proteoglycan synthesis, a qualitative comparison can be made among different loading regimens in different experiments.

The culturing procedure for direct compression is normally a two-step process; the cell-seeded scaffold is left in medium under normal conditions and, subsequently, moved to a machine for mechanical stimulation. Alternatively, a media perfusion setup can be implemented that removes the need for manual feeding, allowing the samples to remain in the bioreactor for the study's duration, decreasing the possibility for contamination [461, 462]. Since native cartilage responds negatively to static loading, devices have often been designed to load a construct dynamically [463]. However, for static loading, it has been postulated that it is not necessarily the mechanical load that is detrimental but the limited diffusion of wastes and nutrients under such conditions, as well as possibly a decrease in the local environment's pH [464]. Whether this is the mechanism or not, static compression has been shown to inhibit matrix secretion [199,321,464–467]. Most studies cite limited mass transport as the main problem with static loading.

Dynamic compression, in which the loading is cyclical, has been shown to a beneficial stimulus when compared to static compression. Individual experiments have used a variety of methods and devices to apply compression, and the results for these studies vary widely, even under the same testing conditions. The main parameters that should be considered for dynamic compression are the frequency of the applied load, the duty cycle, the strain or force used, and the duration of the experiment. Past studies have looked at frequencies ranging from 0.0001 to 3 Hz, strains from 0.1 to 25%, loads from 0.1 to 24 MPa, and durations lasting hours to weeks [199,322,465,468–477]. Some researchers have been constrained in their range of parameters by equipment capabilities; for example, many older devices were unable to produce frequencies above 0.1 Hz. Others problems can exist, such as lift-off of the compression platen from the cartilage sample, making the actual applied strain difficult to measure. Despite these limitations, many informative results have been obtained using a variety of approaches. The combined results give a good view of the benefits of dynamic compression for a wide range of loads and frequencies.

Dynamic, direct compression applied to cartilage explants and/or cell-seeded constructs can induce increases in proteoglycan and collagen synthesis, as well as more robust mechanical properties. Past studies showed that dynamic loading increased ^{35}S-sulfate and ^{3}H-proline incorporation, indicators of proteoglycan and collagen synthesis, respectively, by 15-40% [465,470]. More dramatic results were obtained when dynamic compression was applied in conjunction with a growth factor. For example, proteoglycan and collagen synthesis was increased 180% and 290%, respectively, when the growth factor IGF-I was included during a 0.1 Hz, 3% compression regimen [199]. However, chondrocytes in the tissue were shown to respond differently based on their depth in the tissue. Cells from the deep zone of cartilage explants produced 50% more GAG than static controls under a 1 Hz, 15% strain stimulation, while superficial cells had no significant change. Superficial cells, though, had 40% more ^{3}H-thymidine incorporation than controls, this time with deep cells having no significant change [469]. Direct compression can also affect the magnitude of the aggregate modulus, likely caused by the accumulation of matrix molecules. One such study used a simple, peak-to-peak compressive strain amplitude of 10% at a frequency of 1 Hz with three consecutive 1 hour on/1 hour off cycles per day, 5 days per week for 4 weeks [322]. Using this setup, GAG composition was 33% higher than free-swelling controls. Furthermore, the aggregate modulus was recorded as 100 kPa which is on the same order of magnitude as native articular cartilage (500-800 kPa) [50] after only a month in culture.

Recent studies using dynamic direct compression and TGF-β3 have attained physiological levels for mechanical properties (H_A = 1.3 MPa) and proteoglycan concentrations (8.7% wet weight) [478]. Additionally, these characteristics seem to be uniform throughout the constructs [479]. The latest approaches to direct compression culture take advantage of the synergistic effects between physical and biochemical stimuli, with attention paid to the timing of the application [480–482]. For example, the physiological levels previously mentioned were attained using growth factor treatment for a period of two weeks, followed by dynamic compression for four weeks. The applied strain amplitude was also gradually decreased as matrix was accumulated in the construct. Future work should incorporate knowledge of these time-dependent aspects of the cell response to accelerate the creation of a functional tissue engineered construct.

Mass transfer conditions for a construct under direct compression are slightly better than under static culture. As with other culture systems, construct thickness is limited by the diffusional properties associated with porous scaffolds. Dynamic compression helps alleviate these diffusion limitations through pressure gradients within the construct as well as a secondary mixing effect on the surrounding media. The cells still receive their nutrients through diffusion from the culture media, but transport during compression is enhanced by a dynamic pressure gradient created at the construct's surface [483]. Compression of the scaffold creates a higher hydrostatic pressure at the center of the construct than at the surface, which causes variation in fluid velocities within the construct as the applied load changes. While diffusion of smaller molecules is not affected by the pressure differences created within the scaffold, the movement of larger macromolecules might be hindered [464]. However, recent work has shown that dynamic compression might have the capability to increase the concentration of large solutes within the construct by way of convective transport [484].

4.2   HYDROSTATIC PRESSURE

Chondrocytes in articular cartilage experience hydrostatic pressure during compressive loading of the tissue. The solid matrix of cartilage has a small effective pore size, preventing the rapid flow of fluid out of the cartilage and into the joint space. Therefore, pressure within the tissue cyclically increases during each instance of compressive loading. While it has been shown that this fluid pressurization has a critical role in bearing loads and decreasing surface friction, the effect on chondrocytes within the tissue is less clear. The synovial fluid within the joint capsule transmits pressure to the water trapped within the cartilage's matrix, producing a uniform load on chondrocytes in the tissue. Under physiological levels of hydrostatic pressure (7-10 MPa) [35,485], cartilage is incompressible, resulting in minimal tissue deformation [486]. Loading in this manner is relatively safe for the structural integrity of the tissue since it is a pure hydrostatic pressure that does not stretch or shear the tissue matrix. Early attempts at hydrostatic stimulation used very low pressure changes in a gas phase [487, 488], whose effects may be questionable since the fluid phase may experience no pressure change at all; more recent experiments pressurize the fluid itself which allows much higher magnitudes. Application of hydrostatic pressure has thus far resulted in tissue engineered constructs with aggregate modulus values approaching 300 kPa [386], and its combination with growth factors has shown both additive and synergistic effects in improving construct properties [390].

As with direct compression stimulation, researchers can use either a two-step process for culturing (alternating hydrostatic pressure and non-loaded culture conditions) [489–492] or incorporate a continuous media perfusion approach to minimize handling. For the former, samples are cultured primarily in control, non-loaded conditions but, at prescribed times, are moved to a hydrostatic chamber to undergo loading at the researcher's discretion. This process is repeated throughout the culturing period. The second option incorporates a media perfusion system into a hydrostatic pressure device [493, 494]. This approach has several advantages over the two-step process. The samples do not have to be moved as much, which reduces the possibility of contamination, and the process can be automated to run for long periods of time without any need for human manipulation. Mass transfer is still limited in this setup since fresh media cannot enter the chamber during pressurization [495], but pressurization cycles are usually short enough that this break in perfusion does not affect the nutrient levels significantly.

Constant hydrostatic pressure applied for long periods has been shown to have a negative impact on matrix secretions and cell viability [492, 496, 497]. As seen with other mechanical stimuli, static loads over long periods are not usually beneficial to tissue formation. Even low frequency loading can induce a negative response. Application of a 0.0167 Hz stimulus inhibited sulfate incorporation over short and long durations [492]. It is possible that low frequency stimulation is experienced by cells in a similar manner to static loading. However, a number of studies have also shown positive effects from constant hydrostatic pressure applied for short durations. An increase of 32% in glycosaminoglycan synthesis was observed when a constant pressure of 10 MPa was applied to a high density chondrocyte monolayer for four hours [489]. Recent studies using static pressurization have achieved physiological levels of mechanical and biochemical properties for tissue engineered cartilage constructs [386, 390]. Young bovine chondrocytes formed using the self-assembly process were exposed to 10 MPa of pressure for 1 hr/day for five days, starting at day 10 of a 28 day study. The aggregate and Young's moduli of the constructs were 0.273 and 1.6 MPa, respectively, and glycosaminoglycan and collagen compositions were 6.1% and 10.6% (wet weight), respectively [386].

It has been shown that a window of pressures and frequencies exists between 0.1 and 15 MPa and 0.05 and 1 Hz, respectively, that produces positive results when culturing chondrocytes [489–491, 493, 494, 498, 499]. If hydrostatic pressure exceeds the physiological range, negative changes can occur in the cell, such as decreased matrix synthesis and increased expression of inflammatory cytokines and heat shock proteins [496, 497, 499–504]. High-density chondrocyte monolayers exposed to 10 MPa at 1 Hz for 4 hours a day showed a nine-fold increase in type II collagen mRNA, a 20-fold increase in aggrecan mRNA, and a 65% increase in GAG synthesis [489, 490]. In cell-seeded scaffolds exposed to a 3.5 MPa, intermittent force (5/15 s on/off for 20 min every 4 hr), concentrations of sulfated proteoglycans were twice as high as controls [493]. While intermittent hydrostatic pressure, which mimics physiological conditions (often 1 Hz), has been shown to be beneficial, studies are now emerging where statically applied hydrostatic pressure can also increase construct properties. For example, when comparing the effects of 1, 5, and 10 MPa under static, 0.1 Hz, and 1 Hz conditions, it was found that 10 MPa static HP significantly increased both construct compressive and tensile properties while 10 MPa, 1 Hz treatment only resulted in a significant increase in compressive properties [390].

Not all studies using dynamic hydrostatic pressure induce an increase in the expression of extracellular matrix molecules. In two studies using a 345 kPa, 5/30-sec on/off regimen, sulfated

GAG and collagen secretions decreased when compared to non-loaded controls [493,494]. Hydrostatic pressure has also been shown to increase apoptosis in osteoarthritic chondrocytes [505], as well as cells not surrounded by a pericellular matrix [504]. A protective matrix might be critical to the success of cells exposed to dynamic hydrostatic pressure. Stem cells differentiated in this environment show a rapid accumulation of pericellular matrix in comparison to non-loaded controls [506]. Furthermore, chondrocytic gene expression was enhanced, as was cell viability. Early time points typically show negative results whereas longer culturing times reverse that trend [494,495]. For example, cartilage explants exposed to intermittent pressure showed inhibited proteoglycan synthesis during early loading periods but increased synthesis after 20 hours of loading [492]. This response could be due to a lack of a pericellular matrix in early cultures and the subsequent formation of the surrounding tissue over time.

Many of the variations in studies with hydrostatic pressure can be attributed to donor variability, animal source, and topographical location in the joint [499]. Significant differences in marker incorporation rates can exist among donors and across species. The topographical region of the excised tissue can also affect results because of the varying composition of zonal chondrocytes. However, when combined these results indicate that hydrostatic pressure is important to the maintenance of a chondrocytic phenotype, especially when cells are grown in a three-dimensional environment.

Positive effects have been observed when the culturing conditions of the cells under hydrostatic pressure are modified to alter metabolic activity. Culturing cells in a low concentration of 1% fetal bovine serum at 10 MPa and 1 Hz increased aggrecan expression by 31% and type II collagen expression by 36% [489]. Several recent studies have investigated the effect of oxygen tension on growth and differentiation of chondrocytes. Alterations in pressure can change the amount of oxygen dissolved in the culture media. Many gasses dissolve more readily at high pressures, so the environment in a hydrostatic chamber has to be carefully controlled. Hydrostatic pressure chambers use either a gas or liquid interface to transmit the load to the cultured cell. Both types of devices have advantages and disadvantages. A pure-liquid hydrostatic chamber is the simplest option for keeping dissolved gasses at a known level, and this is the most typical setup currently used. However, chambers using both gas and liquid phases are able to vary the partial pressures present within the culture media. Studies have shown that sulfate incorporation is maximized under atmospheric concentrations (PO_2=21%), but proteoglycan aggregation is maximized at reduced concentrations (PO_2=3-5%) [507]. Reduced PO_2 also stimulates chondrocyte proliferation and type II collagen secretion when using a 30/2-minute on/off loading regimen [491,498].

Hydrostatic pressure stimulates the chondrocytic phenotype of cells grown *in vitro*. Primary chondrocytes that are expanded in monolayer rapidly lose the chondrocytic phenotype [96]. However, stimulation with hydrostatic pressure can help recover the proper expression profile associated with these cells [508–510]. For example, dedifferentiated chondrocytes cultured as cell pellets under a 5 MPa, 0.5 Hz pressure regimen showed a 5-fold increase in aggrecan gene expression and a 4-fold increase in type II collagen gene expression when compared to non-loaded controls [510]. In a separate study, synthesis of chondrocytic proteins was significantly increased for dedifferentiated cells exposed to intermittent hydrostatic pressure [509]. In addition to the biochemical and phenotypical effects, hydrostatic pressure has been shown to influence mechanical properties critical to the function of cartilage constructs. At the counterintuitive frequency of 0 Hz (i.e., static), 10 MPa, applied for 1 hour on days 10-14 of a 4 week culture, was shown to significantly increase aggregate modulus values by 1.4-fold. This regimen also affected functional properties that seem to be difficult to improve upon, namely tensile modulus and strength along with corresponding collagen content, which increased over 2-fold [386]. Hydrostatic pressure has also been shown to act synergistically with growth factors. A combination of 10 MPa static hydrostatic pressure, applied for 1 hour a day for 5 days, and 30 ng/ml TGF-β1 had an additive effect on the mechanical properties, increasing the aggregate modulus by 164% and the Young's modulus by 231%, approaching 300 kPa and 2 MPa, respectively. Additionally, the combined treatment had a synergistic effect on collagen content, increasing it by 173% [390].

Since hydrostatic pressure appears to have a positive effect on the chondrocytic phenotype of cells, recent efforts have used it to help stimulate chondrogenic differentiation in adult stem cells [506, 511–516]. Gene and protein expressions for synovium-derived mesenchymal stem cells were enhanced by intermittent pressurization, showing upregulation of proteoglycan core protein, type II collagen, and SOX-9 [514]. Adipose-derived stem cells expressed a chondrocytic phenotype under hydrostatic pressure and accumulated a pericellular matrix more rapidly than non-loaded controls [506]. A variety of hydrostatic pressure regimens had chondrogenic effects on bone marrow-derived mesenchymal stem cells [511–513, 515]. Furthermore, synergistic effects between growth factors such as TGF-β3 and hydrostatic pressure have been observed for adult stem cells [513]. The loading regimen still plays an important role, though. Low magnitude stimulation has been found to favor SOX-9 and aggrecan expression, whereas high magnitudes favor type II collagen expression and synthesis [512]. These findings might help in determining a more complex loading regimen that targets certain gene expressions at set times.

As with other mechanical stimuli, hydrostatic pressure might assist in organizing cartilage matrix molecules into a more functional structure. Past work has shown that chondrocytes cultured with exogenous chondroitin sulfate formed an abundant cell-associated matrix when exposed to cyclic pressure. Control samples did not incorporate as much chondroitin sulfate and had a less organized matrix when examined by transmission electron microscopy [517]. Self-assembling chondrocyte cultures exhibited higher protein synthesis levels under intermittent pressurization and also showed formation of lacunae surrounding the cells [312]. This structure is similar to that seen in native cartilage and could be critical to the protection of cells in a mechanically-loaded tissue.

4.3   SHEAR BIOREACTORS

Four general categories of shear bioreactors have been investigated for tissue engineering studies. The first is a solid-on-solid, contact shear that attempts to replicate the physiological situation where cartilage rubs against either cartilage or meniscus. The second type, fluid shear, focuses on using fluid flow as a source of shear for monolayer cell populations or cell-seeded constructs and is hypothesized to increase nutrient and waste transfer to increase cell metabolism during culture. The third type of shear, direct fluid perfusion, is a stimulus that has less connection to the physiological conditions in normal joint motion but, instead, was developed primarily to facilitate nutrient transfer through the bulk of a three-dimensional scaffold. The last category, utilizing low shear "microgravity" bioreactors, applies minimal loading to constructs floating in a fluid environment that has flow characteristics that greatly enhance mass transfer to and from the cells.

4.3.1   CONTACT SHEAR

Shear loading is one of several physiological condition that provides mechanical stimulation in normal joint function. While solid-on-solid shearing is minimal because of fluid pressurization in cartilage [7], small amounts of contact shear still exist. The rubbing of two solid materials can have elements of compression and tension that affect the response of cells within the tissue. Bioreactors that replicate this form of mechanical stimulation often are attempting to induce cells to synthesize cartilaginous matrix molecules, as well as trying to create a surface that has frictional properties more akin to the native tissue. As yet, only a few instances of "contact shear" bioreactors have been reported in the literature, likely due to the non-uniform stimulation that is applied through the depth of a sheared construct. Future studies might begin to incorporate contact shear stimulus as a later-stage stimulus once significant matrix has already been deposited in the construct.

Shear bioreactors can be designed to apply translational or rotational strains. In translational shear devices, the construct typically remains fixed to the bottom surface while the top surface moves along one axis [518, 519]. Rotational shear devices apply a small amount of compressive strain and then rotate around the z-axis to produce strains in the construct [520–522]. This is the same motion as used for torsion tests or rheometry although for purposes of stimulation rather than characterization. Alternatives to these traditional approaches do exist, one of which tries to replicate the physiological mechanical environment with a loading shaft that rolls across the tops of fixed constructs, applying a low level of frictional shear (0.5 N normal force) in a cyclic manner [389].

As with direct compression and hydrostatic pressure, dynamically applied shear strains showed more promising results than static conditions. Dynamic shear of 2% at 1 Hz produced constructs with 40% more collagen, 25% more proteoglycan, and 6-fold higher equilibrium modulus [519]. Interestingly, loading duration was minimal – only 6 minutes of shear every other day produced these increases after four weeks. Another research group showed that dynamic shear of 1-3% at 0.01-1 Hz could increase protein synthesis 50% and proteoglycan synthesis 25% [522]. The addition of IGF-I to cultures undergoing shear was found to have a synergistic effect on protein and proteoglycan synthesis that was independent of any improvement of convective diffusion [521]. Applying an interface motion i.e., frictional shear) to cell-seeded scaffolds has been shown to increase cartilage oligomeric matrix protein expression [518]. Future research will investigate the role contact shear has on matrix organization through the depth of engineered constructs.

4.3.2   FLUID SHEAR

While the application of fluid shear forces is more typically associated with vascular tissue engineering, it has been hypothesized that individual chondrocytes might sense shear forces as fluid flows in and out of the solid matrix during compression. Cone viscometers have been used extensively to understand the effects of shear on chondrocyte monolayers. More recent work has focused on using high-shear fluid devices as stimulatory bioreactors or solely as seeding apparatuses.

One of the simplest "bioreactors" is the spinner flask, which uses an impeller to mix oxygen and nutrients throughout the media. Cell-seeded scaffolds are fixed firmly inside the flask away from the impeller. The samples benefit from increased nutrient and waste transfer, as well as experience controllable levels of shear. The container shape and mixing rate can both affect the shear patterns throughout the culture environment, leading some groups to investigate close alternatives to the spinner flask, such as the wavy-walled bioreactor [523, 524]. In all of these fluid shear bioreactors, cells can either be seeded onto scaffolds before they are inserted into the flask or inoculated directly into the media, gradually attaching to scaffolds already in the flask [525]. Much work has gone into optimizing cell seeding of scaffolds using this technique, which will be discussed later in this chapter. Another variation of the mechanically-stirred environment is the orbital shaker/rotating plate, which can slowly mix media in a culture without much turbulence [526].

Cell-seeded scaffolds cultured in spinner flasks have shown both positive and negative results, depending on the level of shear seen by the cells. In one study, cartilage constructs from fluid-sheared cultures (at 50 rpm) were more regular in shape and contained up to 70% more cells, 60% more sulfated glycosaminoglycan, and 125% more total collagen [527]. The increase in matrix constituents is likely due to the larger number of cells in the constructs compared to controls. While the improved matrix composition is a plus, there are also undesirable side effects associated with high shear systems. Cell damage has been observed at 150-300 rpm in microcarrier cultures [528], and although there is no apparent physical cell damage at 50 rpm, a fibrous capsule does form on the construct surface [527]. While fibrous encapsulation does occur in most systems because of increased nutrient availability at the construct surface, its presence could also indicate a protective response to shear forces. The local shear force experienced by the cells is produced by eddies created by the turbulent flow of the impeller. Cell flattening, proliferation, and formation of an outer capsule is caused by the pressure and velocity fluctuations associated with turbulent mixing [527]. Other experiments have also seen increases in total collagen content for constructs cultured in spinner flasks [529]. However, a large percentage is likely type I collagen since that is what composes the capsule surrounding the construct. The mixing rate has a limited effect on the amount of proteins secreted by cells but does affect what types of proteins are made and whether they are incorporated into the construct. Cell-seeded scaffolds exposed to any intensity of mixing (80-160 rpm) synthesized more collagen and GAG than controls but actually retained lower fractions of GAG within the scaffold [523]. This loss of GAG from the construct is caused by the continual convective flow in the spinner flask.

Spinner flasks and other impeller-based bioreactors are popular because they increase the mass transfer rate to the cells. However, forming hyaline tissue via an impeller bioreactor is difficult. Non-uniform mass transfer rates, nutrient and pH gradients, and shear gradients, which cause a non-uniform mechanical stimulus over the sample, all contribute to inferior tissue formation compared to other bioreactors [354]. Shear force at the surface of the impeller is ten times higher than anywhere else within the bioreactor [530]. Because of this, samples closer to the impeller could experience injurious levels of shear while samples further away might not be stimulated at all. If positioned correctly, fibrous capsulation of the construct will be minimal, and the cells will still benefit from enhanced nutrient transfer. However, low mixing rates or large distances from the shear source could decrease mass transfer, creating a stagnant environment with increased pH caused by lessened mixing. For successful use of a stirring bioreactor, a balance has to be obtained between the level of shear force and the extent of nutrient and oxygen transfer in the media.

Some of the limitations mentioned above have been remedied by modifying the fluid shear devices to lessen the magnitude of shear and create a more homogenous flow environment. The cone viscometer, which consists of a small-angled cone that rotates in media above a flat surface seeded with cells, can achieve a uniform shear distribution with values ranging from 10^{-3} to 10 Pa [531]. This type of device has been attractive to researchers because it can apply a laminar shear force at a constant, controllable level. High-density chondrocyte monolayers exposed to a 1.6 Pa shear force showed a 2-fold increase in GAG synthesis but also a 10- to 20-fold increase in prostaglandin E_2 release and 9-fold increase in tissue inhibitor of metalloproteinase mRNA, both of which indicate an inflammatory response to the fluid shear [252]. Additionally, interleukin-6 and nitric oxide levels, which are reliable indicators of osteoarthritis, increased due to this type of mechanical stimulation, and chondrocytic gene expression (aggrecan, type II collagen) decreased significantly [532, 533]. These results show that fluid shear, at least when applied to chondrocytes not in a scaffold, might not be beneficial to chondrogenesis.

Mechanically stirred bioreactors might not be optimal for growing hyaline cartilage, but they do function well for attaching cells to fibrous mesh scaffolds. Spinner flasks are one of the most efficient cell-seeding techniques when using pre-formed scaffolds [525]. Mixing provides for rapid, high yield attachment and a more uniform distribution of cells throughout the scaffold, as well as inducing better overall matrix production in the construct. Static seeding results in cells located primarily in the lower half of the construct while dynamic seeding distributes cells more evenly throughout the scaffold [529]. If the scaffold material is coated with protein (fibronectin, collagen, etc.), then dynamic seeding can produce an even higher yield of attachment as well as increasing migration of the cells into the scaffold [423]. Successful tissue engineering results rely heavily on well-seeded scaffolds, and dynamic seeding provides a relatively simple approach for obtaining high cell density constructs. However, some polymers, like hydrogels, do not need to be seeded in this manner and can be distributed evenly at high densities without the use of a mechanically stirred bioreactor.

4.3.3   PERFUSION BIOREACTORS

An alternative type of shear bioreactor is one that incorporates fluid perfusion. Devices using this approach are designed to continuously flow media either around or through (direct flow perfusion) a porous scaffold populated by cells. The fluid flow can be controlled to apply a range of shear forces while enhancing the availability of nutrients continually during culture. Direct perfusion systems have been used extensively for bone tissue engineering applications, and experiments conducted for cartilage appear promising.

One of the simplest approaches to fluid perfusion is to flow media steadily though a chamber containing cells or cell-seeded scaffolds. Some bioreactors use this type of perfusion to feed the cells continuously while applying other mechanical stimuli like direct compression or hydrostatic pressure. Perfusion bioreactors can have both positive and negative effects on tissue growth. In one experiment, engineered cartilage constructs in static culture accumulated 300% more sulfated glycosaminoglycans, incorporated 180% more ^{35}S-sulfate, and expressed aggrecan and type II collagen 350% and 240% greater, respectively, than perfused constructs [534]. This result could be caused by a loss of newly synthesized matrix molecules to convective flow, which is not present in static conditions. Culture time can also affect the overall response of constructs to direct perfusion. Glycosaminoglycan synthesis and retention are inhibited at early time points. However, with extended culture, matrix accumulates in the scaffold and total proteoglycan amounts are greater than non-perfused controls [535]. Another experiment showed that continuous media perfusion to a culture dish can increase matrix production 50-70%, similar to "through-thickness" perfusion experiments, which are discussed below [536]. This increase in synthesis could be caused by improved mass transport or shear force stimulation. However, if the cells are shielded from the surrounding flow, then the sole effect is increased nutrient availability. Experiments using this setup have been used for growing cartilaginous tissue. For example, agarose-encapsulated scaffolds placed in a chamber with continuous media perfusion provided cells with nutrients via diffusion, similar to the physiological environment, but also shielded them from shear forces [537]. High flow rates allow for high nutrient concentrations, but the drawback is increased shear forces that might produce a negative cell response as discussed previously. The next section will discuss bioreactors that use this perfusion approach in "low-flow" systems designed to enhance nutrient transfer without any mechanical stimulation.

For three-dimensional constructs, a bioreactor that forces media *through* the scaffold gives the most thickness-independent results. This type of system is called a direct perfusion bioreactor. Cells throughout the construct experience fluid shear as media moves through the scaffold structure. In response to this stimulation, cells secrete extracellular matrix. It should be noted, however, that the matrix molecules resulting from this shear effect are not necessarily those characterizing hyaline cartilage. As seen in previous experiments with high shear, fibrous matrix composed of type I collagen can dominate the construct.

Direct perfusion bioreactors also effectively align cells in the direction of flow, which can be advantageous when producing a tissue with specific cellular orientations like articular cartilage [538]. However, not all cells in articular cartilage are aligned in the same direction, so this effect may or may not be advantageous for creating a similar structure to the native tissue. Variations exist in direct perfusion designs, but one commonality is a tight fit between the scaffold and walls of the media chamber. If the scaffold has space around it, less fluid is forced through the pores and uniform mechanical effects are not achieved. Basically, the chamber becomes a perfusion system and not a direct perfusion system. A major benefit of this bioreactor is the continuous influx of fresh media to cells through the thickness of a scaffold. Additionally, perfusion removes the need for manual media changes, decreasing labor and reducing the risk of contamination.

Modifications can be made to direct perfusion bioreactors to alter the growth environment of the samples. For example, media that has run through the system can easily be mixed in various proportions with fresh media. Recycling some of the culture media keeps beneficial proteins secreted endogenously by the cells i.e., growth factors, matrix molecules) in the system. Another possible modification to the system involves controlling gas concentrations in the fluid. By adjusting the oxygen content of the media, researchers can vary the concentration exposed to the cells [539]. If the tissue becomes denser, more oxygen and nutrients can be added to compensate for the increased oxygen usage. Experiments with variable levels of oxygen tension can also be easily controlled in a perfusion system.

Direct perfusion of tissue engineered constructs can affect cell proliferation and viability, matrix secretion, and tissue uniformity. Direct perfusion bioreactors with limited levels of shear (< 0.01 Pa) can stimulate cell proliferation and increase the production of proteoglycans and collagen [540, 541]. Cell-seeded scaffolds cultured in a direct perfusion bioreactor running at 1 μm/sec (flow rate of 7.6 μL/min) for four weeks showed an increase of 184% in glycosaminoglycans, 155% in ^{3}H-proline incorporation, and 118% in DNA content [538]. These increases are promising although secreted molecules associated with injury response were not measured. Another research group looked at applying direct perfusion at a higher linear velocity of 10.9 μm/sec (flow rate of 50 μL/min) [459]. The resulting constructs were composed of 25% (dry weight) glycosaminoglycans and 15% (dry weight) type II collagen (the balance being non-degraded polymer). While the collagen composition is still significantly lower than native cartilage levels (50-73%), the absence of type I collagen indicates that direct perfusion might be a possible option for growing hyaline cartilage. Unfortunately, tissue growth in the scaffold was non-uniform, with more matrix deposition observed for the scaffold side facing fluid flow. Increasing the flow rate could mediate this problem since the energy of the fluid flow dissipates as it passes through the scaffold, stimulating cells less as it progresses through the scaffold. However, increased shear is likely to affect the cells at the surface negatively. Finding a balance between complete perfusion and low shear forces is necessary if direct perfusion bioreactors are to be used for cartilage tissue engineering. Rotating bioreactors, discussed later in this chapter, are a possible solution to this problem.

Significant problems exist with using direct perfusion bioreactors for cartilage engineering. One is that cellular secretions are non-uniform through the thickness of the construct. Since fluid flows from one side of a scaffold to the other, the front surface experiences greater mechanical stress due to the oncoming flow. Conversely, the back surface only experiences the shear stress inside its pore structure and not on its face. This flow environment results in a construct that has a thick matrix layer on one side and minimal matrix on the other. Additionally, matrix secretion in the bulk of the scaffold is non-uniform since the energy associated with fluid flow either dissipates or concentrates in regions according to the scaffold structure. Another problem is the induction of molecules associated with injury response rather than matrix formation. Studies have shown that shear levels as low as 0.092 Pa (0.92 dyne/cm^2) can have adverse effect on cells [542]. While chondrocytes are considered robust when exposed to mechanical stress, studies have shown that turbulent flow can produce a negative effect even on chondrocytes [527–529]. The cells might not die, but their protein secretions do become phenotypically altered, resulting in a deposited matrix that is mechanically inferior to native cartilage. In this case, chondrocytes produce a thick, fibrous matrix composed mainly of type I collagen that effectively isolates the cells from the turbulent flow [543]. High-shear direct perfusion devices induce a fibrous response similar to the capsule formed in some spinner flask cultures, although it is usually restricted to one side of the construct.

4.3.4   "LOW-SHEAR" BIOREACTORS

Flow-based bioreactors are attractive systems for tissue engineering because they improve mass transfer rates, effectively increasing nutrient concentrations and decreasing waste levels in the culture environment. While high-shear perfusion can successfully stimulate matrix production, the resulting tissue is typically fibrous in nature rather than hyaline. Slower fluid flow rates are hypothesized to have a general stimulatory effect on matrix synthesis while still allowing cells to express a chondrocytic phenotype. This is the premise behind low-shear, rotating bioreactors.

Some of the most successful literature reports for cartilage bioreactors come from a modified version of the clinostat, which was first described in 1872 by Julius von Sachs [544]. Its modern day representation, the rotating wall bioreactor, provides a culture environment in which constructs are continuously suspended in media. Sometimes described as a "microgravity" environment, this device was developed by researchers to investigate the effect of free-fall on cell and tissue growth [545]. Within the past decade, rotating bioreactors have found success as a low-shear, high diffusion bioreactor for many cell types. The ability of perfusion bioreactors to provide nutrient-rich environments for cells in a stimulatory environment was carried over into the design of this more sophisticated device. The original design is comprised of a media-filled, cylindrical vessel that rotates around a central axis (also capable of rotation) at 15-30 rpm, which keeps constructs or cells floating in suspension. Rotation speed has to be adjusted throughout the culture period to balance any gravity effects on the samples. Gas exchange occurs through a gas permeable membrane that forms a hollow, inner cylinder. Dynamic laminar flow in rotating bioreactors provides efficient oxygen supply and allows newly synthesized macromolecules to be retained in the developing constructs [546]. Later versions of the rotating bioreactor have modified the shape of the vessel and the mechanism for gas and media perfusion. The culture environment present in rotating bioreactors make it attractive for not only tissue engineering studies, but also more basic studies focusing on cartilage healing and cell aggregation [547, 548]. The major difference between rotating bioreactors and past perfusion systems was a reduction in shear force, since high or even moderate levels of shear are undesirable in the formation of hyaline cartilage. A rotating fluid environment was found to be the best way to produce a low-shear, high-mass transfer bioreactor [460, 549–552].

The rotating bioreactor is capable of adjusting shear levels associated with fluid/construct interaction because of its unique design. Shear forces can induce either positive or negative responses from cultured cells, and a threshold level of 0.092 Pa seems to demarcate that point for rotating bioreactor systems [542]. Constructs cultured in this environment remain suspended in the media by two forces: gravity and fluid flow. Samples 'fall' through the media while rotating fluid flow acts in the upward direction, keeping the samples suspended but also exerting a slight shear force. Altering the rotation rate of the vessel can effect different flow lines and shear environments around the sample. Low-shear environments are typically produced by slowly rotating both the inner and outer cylinders at nearly the same rate. Initial experiments used cell-seeded microcarriers to investigate the effect of "microgravity" on cell growth and development [542, 552, 553]. Over time in culture, the microcarriers slowly aggregated to form larger cell-matrix constructs [543, 554]. Subsequent experiments with larger constructs showed that the samples tended to remain near the ends of the media chamber, not distributing as widely as smaller particles. Because of this, newer versions of the rotating bioreactor (Synthecon, Inc.) have altered the aspect ratio of the media chamber, producing an environment more conducive to culturing large constructs [554]. The culture environment is not ideal, though, since the flow patterns inside the bioreactor tend to slowly tumble large constructs through the media; an action that introduces higher shear levels caused by turbulent fluid motion across the construct surface [553]. Small-amplitude, long-period oscillations in the fluid-wake at this interface may be the source of mechanical stimuli felt by the cells [549]. Additionally, the magnitude and direction of shear on the construct constantly changes, which might be good (dynamic mechanical stimulus) or bad (un-definable forces). The stress exerted on a construct in a bioreactor rotating at 19 rpm was calculated to be $\sim$0.15 Pa (1.5 dyne/cm^2) [549], which is significantly higher than the shear level of $\sim$0.0005 Pa (0.005 dyne/cm^2) measured for microcarrier beads in the same environment [552]. However, this shear stress is still significantly lower than many other fluid flow bioreactors.

Published results show wide use of rotating bioreactors for cartilage tissue engineering. Past findings have shown that rotating bioreactors produce higher fractions of glycosaminoglycans and collagen than mixed flasks or static culture [353, 551]. Constructs cultured for six weeks produced tis- sue that had glycosaminoglycan and total collagen compositions that were 68% and 33%, respectively, of native cartilage levels [353]. Similar results were obtained in a subsequent study, with engineered constructs accumulating 75% of native glycosaminoglycan and 39% of native type II collagen compo- sitions [551]. Additionally, extending culture to seven months increased glycosaminoglycan content beyond physiological levels although collagen remained at 39%. The accumulation of matrix also affected the mechanical properties, with equilibrium moduli (950 kPa) and hydraulic permeability (5×10^{-15} m^4/N-s) reaching values comparable to healthy cartilage.

Improvements to the engineered tissue are not limited to increased matrix production. The morphology of the constructs shows more uniform deposition of matrix than in other bioreactors [353, 550]. Collagen and proteoglycan accumulation occurs in both the peripheral and central regions of the scaffold, and fibrous encapsulation is minimal or even non-existent [555]. These results indicate that oxygen and nutrients reach the construct center in sufficient amounts, which is important for growing large tissues. While low oxygen concentrations are more representative of the physiological environment in articular cartilage, anaerobic conditions have been shown to cause poor matrix production [556, 557]. The mass transfer enhancements of the rotating bioreactor are, therefore, critical to its success. Oxygen and nutrients move further into the scaffold, facilitating the growth of constructs as thick as 5 mm after 40 days of *in vitro* culture [353]. The culture system has also been shown to increase cell proliferation/viability and decrease nitric oxide production, both of which indicate a good growth environment [558].

Bioreactors for cartilage tissue engineering should provide an environment that is conducive to retaining the chondrocytic phenotype. A number of studies have investigated rotating bioreactors, or modified versions of these devices, for their capability to re-differentiate chondrocytes that have non-ideal expression patterns [559–562]. De-differentiated chondrocytes transfected with BMP-2 were cultured in a "rotating shaft" bioreactor to induce chondrogenic changes in gene and protein expressions, as well as stimulate rapid matrix accumulation [559]. Results in static culture were inferior to those in the bioreactor as were results using non-transfected cells. This indicates that a synergistic relationship exists between biochemical and mechanical stimuli, which can be facilitated for cartilage growth by the rotating bioreactor. Constructs grown for three weeks *in vitro* and then implanted *in vivo* for eight weeks showed good histological characteristics and integration with surrounding tissue [560]. Another research group also found the rotating bioreactor conducive to stimulating the chondrocytic phenotype [561, 562]. Cells from aged subjects (~84 years old) were inoculated into the bioreactor without a scaffold and analyzed after twelve weeks in culture. The resulting constructs formed a cartilaginous matrix that was rich in collagen type II and proteoglycans.

The preferred cell type for early cartilage tissue engineering studies was the chondrocyte. However, difficulty with obtaining healthy chondrocytes from patients has driven interest towards other cell types, like stem/progenitor cells (See Chapter 5). Many different types of cells have been used in the rotating bioreactor because of its apparent conduciveness to the chondrocytic phenotype. Bone marrow-derived mesenchymal cells have been used successfully to create cartilage and osteochondral constructs exhibiting good protein accumulation over weeks culture [563–565]. These cells have been characterized as being more metabolically active than either static or simple perfusion environments [566]. Another progenitor cell type, synovium-derived stem cells, has shown an ability to secrete matrix rich in glycosaminoglycans and collagen type II after a month in the rotating bioreactor [567, 568]. Several other cell types including amniotic mesenchymal cells [569], umbilical cord blood cells [570], and embryonic stem cells [571] have all successfully differentiated down the chondrogenic lineage when cultured in the rotating bioreactor. Numerous researchers are actively investigating how the unique conditions of this culture environment can be used to exploit the multi-potentiality of stem/progenitor cells.

Not all bioreactors can accommodate the wide variety of scaffold materials that are present in the field of tissue engineering. The physical, mechanical, and material characteristics of a scaffold can prevent its use in devices that apply large forces or require excessive handling. Rotating bioreactors, however, provide a gentle culture environment that is conducive to many different carriers. Mesh scaffolds, hydrogels, and microcarrier beads can all be cultured with the same ease. Cells can even be cultured without a scaffold, with aggregation occurring within days to weeks. This is not to say that the scaffold-bioreactor compatibility can be totally ignored. In fact, this interaction is incredibly important when determining the shear forces and nutrient diffusion present in the culture environment [572]. Furthermore, the rotating bioreactor cannot overcome disadvantages associated with certain scaffold types or culture parameters, such as low initial cell densities [573]. However, successful experiments have been carried out with a variety of scaffolds. Chondrocyte-seeded poly(DL-lactic-co-glycolic acid) sponges showed formation of hyaline-like tissue when cultured in a chondrogenic media for four weeks [574]. Chitosan scaffolds have also been used in the rotating bioreactor although tissue growth was shown to be strongly influenced by the microstructure of the scaffold [575, 576]. More recently, a chitosan-hyaluronan hybrid scaffold was investigated in this culture environment with results showing near-physiological levels for matrix composition and mechanical properties [577].

A major concern with rotating bioreactors is the random motion of scaffolds in the culture vessel. Researchers typically put multiple constructs in a single bioreactor, which results in groups of tumbling samples that can hit one another or the walls containing them. These unpredictable contacts can kill cells in localized areas or damage the scaffold during early culture periods. Another difficulty is identifying the flow patterns within a bioreactor filled with constructs. One attempt to localize the nutrient flow and keep a more stable culturing environment is the hydrodynamic focusing bioreactor (HFB), which was created by NASA for use in no-gravity cell culturing [543, 553, 554]. As with other rotating bioreactors, the inner and outer walls rotate to produce a range of shear forces. However, instead of having a cylindrical shape, the HFB is a dome. This modification is proposed to focus cells and nutrients together to enhance mass transfer. Another version of the rotating bioreactor is the "rotating shaft" bioreactor [578]. This device uses the motion of the inner cylinder to move attached samples in a continual rotary motion around the central axis. However, the culture vessel is only half-filled with media, so samples move in and out of liquid and fluid phases. This is proposed to increase oxygenation as well as provided slightly higher levels of shear. As with other studies involving rotating bioreactors, experimental results with this device were generally successful for cartilage tissue engineering [559, 560].

4.4   HYBRID BIOREACTORS

As the field of tissue engineering matures, more complex bioreactors have been developed to more faithfully replicate the native environment of target tissues. Each bioreactor reviewed in the previous sections has advantages and disadvantages to its design. Some systems appear to have greater effects on collagen production i.e., direct compression) while others enhance proteoglycan synthesis i.e., hydrostatic pressure). More information is continually being accrued that helps elucidate the complicated relationship between mechanical stimuli and cell response. Successfully applying mechanical stimulation can be difficult because each bioreactor has to be optimized to take advantage of its benefits while minimizing its deficiencies. One possible solution is to combine two or more bioreactors that complement each other's strengths and weaknesses.

Researchers have already begun integrating different mechanical stimuli into single, hybrid bioreactors to elicit a better response from cells. Initial experiments investigating the combination of small axial compression with rotational shear has shown a stimulatory effect on protein and proteoglycan synthesis [520–522]. More complicated experiments involving larger axial strains and sequential periods of stimulation have yet to be performed. An alternative to this approach is an innovative device that dynamically rolls a cylinder across fixed samples to apply both compressive and frictional shear loading [389]. This stimulus upregulated chondrocytic gene expressions and increased protein synthesis in explants after four days of culture. The only other hybrid bioreactor reported for cartilage engineering is one that combines hydrostatic pressure and direct fluid perfusion [579, 580]. This device functions by pressurizing the media as it flows through the interstitial spaces of a cultured construct. Future work will investigate the combined effects of these stimuli on matrix production and tissue organization.

Bioreactors do not necessarily need to be combined into one device to be effective. One of the more promising approaches in cartilage engineering is to use a dynamic seeding environment initially, followed by a different type of mechanical bioreactor for the duration of culture. For example, scaffolds can be seeded and stabilized for a short period in a spinner flask. Then they can be transferred to another bioreactor, such as intermittent pressurization, to facilitate tissue growth. This approach has been found to achieve better results than either mechanical stimulus alone [581]. When cell-seeded scaffolds were cultured for two weeks in a spinner flask and four weeks in hydrostatic pressure, constructs had 3.5 times more GAG and seven times more collagen than static controls. A large field of study is currently open that requires investigation of which specific combinations of mechanical stimuli, as well as their sequences and durations, can produce the best response from cultured constructs.

New bioreactors can be developed that combine the beneficial aspects of several devices into one package. For example, hydrostatic pressure could be combined with a rotating bioreactor to create a stimulating environment that is self-contained. Since the scaffolds are already cultured in a fluid medium, hydrostatic pressure could be applied without removing anything from the sterile environment. This reduces the chance of contamination as well as limiting the amount of man-hours needed to transfer the scaffolds between a stimulation device and a culturing environment. Another possible device would combine direct compression and direct fluid perfusion. This bioreactor could enhance mass transfer while also applying a physiologically-relevant strain with cyclic compression. The difficulty with bioreactor design is the extensive testing and validation required before widespread adaptation. While combining two or more established mechanical stimuli seems straightforward, the biological response of an engineered construct could be totally unpredictable.

4.5   CHAPTER CONCEPTS

- Bioreactors can be used to apply chemical and nutrient gradients in the culture of articular cartilage constructs; it can also be used to apply mechanical forces to the developing cartilage.

- Direct compression, hydrostatic pressure, high and low shear, and hybrid bioreactors have all been examined, oftentimes in an effort to replicate physiological loading regimens.

- Cyclic direct compression has been shown to be more beneficial than static compression. For hydrostatic pressure, however, it has been shown that both static and dynamic applications can result in functional improvements.

- For hydrostatic pressure, it has also been shown that an optimal window in time exists in the application of this force.

- In self-assembly of articular cartilage, 10 MPa of hydrostatic pressure, applied statically for one hour a day, five days total, has been shown to increase functional properties, both compressive and tensile.

- Direct compression and hydrostatic pressure have both been applied in combination with growth factors to show enhanced effects.

- Shear can be applied as either contact or as fluid shear in bioreactors. Translation or rotation shear can be applied directly to constructs. It is hypothesized that chondrocytes experience shear as the tissue is deformed during loading.

- Direct fluid perfusion is another way to apply shear to chondrocytes while increasing nutrient/waste transport.

- As proinflammatory mediators have been seen with the application of shear on chondrocytes, low shear bioreactors have been developed to aid in nutrient transfer without eliciting inflammatory and catabolic factors.

- Several bioreactors can be used at once and at different times. For instance, a spinner flask bioreactor can be used for seeding cells onto a scaffold, and then the construct can be transferred to a rotating wall bioreactor to experience minimal fluid shear with increased nutrient transportation, and, finally, the construct can experience direct compression or hydrostatic pressure in separate bioreactors.

CHAPTER 5

Future Directions

Much progress has been made in the areas of understanding cartilage physiology, development, and pathology, and tissue engineering efforts have made significant strides in employing biomaterials, biochemical agents, and bioreactors. Engineered cartilage as clinical therapy is still seldom seen due to several challenges. Autologous, differentiated chondrocytes represent a limited donor source while allogeneic and xenogeneic sources can have issues associated with disease transmission or immunogenicity. Researchers are also developing methods to integrate new cartilage to existing cartilage and bone. While there is a set of assays often seen in the literature in evaluating cartilage, there is not yet a collection of standard methods that allow cross-platform comparison of the tissue engineered cartilages. Finally, the translation of cartilage therapies to the bedside will require extensive testing for their efficacy and safety. A brief presentation on how the FDA approval process can inform the types of cartilage therapies a company chooses to develop will conclude this chapter.

5.1   CELL SOURCES FOR THE FUTURE

Tissue engineering studies of cartilage have been initiated with differentiated cells specific to articular cartilage. For patients with already diseased or missing tissues, alternative cell sources must be considered. This section gives an overview on the many cell sources currently available: differentiated chondrocytes (of auto, allo-, and xenogeneic in nature), autologous progenitor/stem cell populations, such as mesenchymal, adipose, and skin derived stem cells [299–301, 582, 583], and adult cells of other differentiated lineages that have also been investigated [301, 582]. Other stem cells with potential for cartilage tissue engineering include embryonic and induced pluripotent stem cells.

5.1.1   A NEED FOR ALTERNATIVE CELL SOURCES

A major problem with *in vitro* approaches to cartilage tissue engineering is that the use of native chondrocytes is not currently a practical solution for patient therapies [584]. There are simply too few chondrocytes in the body that can be reasonably harvested to support the generation of tissue constructs that can effectively treat clinically relevant cartilage pathology. Although expansion of these cells is possible, the expanded cells lose their phenotype with each passage [96, 585, 586]. Stem cells have emerged as a possible solution. Particularly attractive in using these cells is the fact that they can be expanded *in vitro* to the required number of cells and subsequently differentiated into the desired cell type. Stem cells can be found in many parts of the body and at many stages of development. Pertaining to cartilage, a vast amount of work has focused on mesenchymal stem cells (MSCs), adult stem cells that can be found in the bone marrow. Other stem cells also have been investigated for their chondrogenic potential, such as adipose derived stem cells [587], embryonic stem (ES) cells [588–593], embryonic germ cells [594], progenitor cells from the placenta [595], and umbilical cord blood stem cells [570].

When considering which class of stem cells to use for therapeutic applications, it is important to understand their basic differences. The first major difference is the differentiation capacity of each. Embryonic stem cells are termed pluripotent, as they hold the ability to become any of the three germ layers. This has been demonstrated with each of the NIH-approved hESC lines, including H9 [596] and BG01V [597, 598]. Efforts into the directed differentiation of each of these cell lines into cells or tissues with therapeutic potential have been pursued for musculoskeletal [599] and neural tissues [598], among others. The pluripotency of these stem cells largely embodies the excitement and danger of using these cells as a therapy. While they can theoretically differentiate into any cell in the body, they also may form tumors called teratomas. Other stem cells, such as MSCs, adipose-derived stem cells, embryonic germ cells, and umbilical cord blood stem cells do not form tumors, but they also do not have the differentiation potential of ESC. Each of these classes of stem cells has been shown to have the capacity to differentiate into cells that produce cartilaginous matrix [570, 587, 594]. In contrast to other stem cells, it appears that ESC are immortal, meaning that theoretically they can be expanded without losing their phenotype. However, the culture of hESCs remains an issue because of the use of feeder layers, such as mouse embryonic fibroblasts (MEFs). These feeders or media conditioned by the feeder cells have been generally used for the culture of hESCs in an undifferentiated state. This co-culture system involving human and animal cells presents practical problems for future therapies since animal products or pathogens can be transmitted. Efforts are underway to address this [600–603] although no alternative to MEFs has been universally adopted. As indicated in their nomenclature, the source of the stem cells is also a major difference, and the successful isolation of each type of stem cell varies. For example, MSCs constitute a low proportion of bone marrow stromal cells and, additionally, may contain genetic abnormalities, caused by exposure to metabolic toxins and errors in DNA replication accumulated during the course of a lifetime [584]. ESC, on the other hand, have been shown to be homogenous and genetically stable in culture.

5.1.2   CHONDROGENIC DIFFERENTIATION OF MSCS AND OTHER ADULT CELL SOURCES

Regarding the chondrogenic differentiation of stem cells, most information gathered centers around MSCs. In addition to cartilage, these are the progenitors of multiple other tissue lineages, including bone, muscle, and fat. In the design of composite structures, such as the articular condyle, MSCs have been used to engineer osteochondral constructs [604–606]. For *in vitro* work such as this, as well as *in vivo* work with stem cells, investigations into their ability to chondrogenically differentiate is commonly defined as a process that results in cells with the ability to produce both collagen type II and glycosaminoglycans [588, 591, 607, 608]. Differentiation factors, such as TGF-β1, BMP-2, and IGF-I, have been used to direct the differentiation of MSCs to a lineage of cells that can secrete cartilaginous proteins [607, 609–611]. Genetic manipulation has been used to induce MSCs to produce growth factors [610]. Others have genetically modified MSCs to express key signaling and transcription factors of cartilage to investigate their ability to help regenerate cartilage [612, 613]. Recently, mechanical stimulation has also been used in directing MSCs to a chondrogenic linage [475,513]. Despite this progress, it remains to be seen whether the pursuits with MSCs will demonstrate the generation of tissue that has the biomechanical wherewithal of native cartilage, or that MSC-derived cartilage can provide long-term solutions to cartilage pathology [614].

Knowledge gleaned from the differentiation of MSCs has been applied to other adult stem cells. Similar chondrogenic differentiation of adipose derived stem cells [615] have been performed with TGF-β1 and BMP-2 [616], and, recently, the role of hydrostatic pressure has also been implicated [506]. Hydrostatic pressure applied at 0-0.5 MPa and 0.5 Hz resulted in a higher rate of matrix accumulation than controls. Adipose stem cells are attractive because they are relatively easier to obtain than MSCs. An even less invasive cell source would be the derivation of multipotent dermal precursors [617,618]. Chondrodifferentiation of dermis-derived cells has been seen by seeding these cells onto demineralized bone matrix [619] and in combination with growth factors [620], as well as using a surface coated with aggrecan [582]. Subsequent purification of the starting skin cell population has yielded tissue engineered constructs that stain throughout for collagen type II with absence of collagen type I staining [301] and improved collagen type II expression over un-purified cells.

5.1.3   CHONDROGENIC DIFFERENTIATION OF ESC

The evidence remains scarce regarding the use of ESC for cartilage tissue engineering strategies. Much of the evidence supporting the use of ESC for cartilage tissue engineering comes from work with mouse ESC [608]. The chondrogenic differentiation of these cells has been demonstrated *in vitro* using BMP-2 (2 ng/ml; 10 ng/ml) and BMP-4 (10 ng/ml) [591]. Their phenotypic stability in a differentiated state has also been investigated [591, 593]. Others have also been able to differentiate ESC into a chondrogenic lineage with the use of special culture conditions with growth factors [588,590], without growth factors [592], and in co-culture with limb bud progenitor cells [589]. An example is the use of hydrogels with mouse ESC that were chondrogenically differentiated with TGF-β1 or BMP-2 [621]. Recently, hESCs were differentiated into mesenchymal precursors, which can be subsequently chondrogenically differentiated with TGF-β3 (10 ng/ml) [599]. For instance, ESC can be exposed to TGF-β3 (10 ng/ml) to form hyaline-like cartilage through a mesodermal lineage [588]. Recently, the use of differentiated hESCs for tissue engineering has been pursued in a modular approach. That is, cells are differentiated as a first step and then, after dissociation and/or purification, assembled into tissue engineered constructs [308]. Since past studies have shown that culturing stem cells under serum-free conditions may result in a lower mitotic index for cells, apoptosis, and poor adhesion [622,623], these studies are particularly notable since the differentiation is performed in serum-free conditions. Another study has demonstrated the musculoskeletal differentiation of human embryonic germ cells; here, a chemically defined chondrogenic differentiation medium with 1% serum and with one of two differentiation factors, BMP-2 and TGF-β3, was used [594]. Aside from the use of growth factors, another tool that has been shown to improve hESC differentiation to a chondrogenic phenotype includes hypoxia [309].

The use of ESC brings about their own challenges. The introduction of serum for *in vitro* applications increases the variability of components in the culture media. However, there are many interactions between individual growth factors and serum components that cannot be ignored. Due to the interrelated mechanisms of growth factor action, in order to study the effect of differentiation factors, it may be necessary for serum (which includes several types of growth factors) to be present. Chondrogenic effects of individual growth factors were demonstrated with levels as high as 20% serum [591, 593]. One has to be cognizant, however, that when serum is used, there is a risk of saturating the experiment with growth factors, yielding results that suggest the growth factor of interest has had no effect compared to controls. This limitation can perhaps be addressed by using a minimal amount of serum [624]. Nonetheless, for clinical applications of stem cells, it is important that protocols exclude the use of animal or human products, like MEFs or serum, that may carry pathogens or potentially increase the antigenicity of the transplanted cells [584]. The elimination of serum, as noted above, has been pursued, and the elimination of murine feeder cells in the media is also being considered [600–603].

5.1.4 IPSCS AND CARTILAGE TISSUE ENGINEERING

Induced pluripotent stem cells (iPSCs) are somatic cells reprogrammed into an ESC-like state through the ectopic expression of the transcription factors Oct4, Sox2 and either c-Myc and Klf4 or Lin28 and Nanog [625, 626]. These cells exhibit the same morphology, proliferation, normal karyotypes, telomerase activity, surface markers, expression of pluripotency genes, and teratoma formation as ESCs. However, given the emerging evidence for variation in differentiation potential between hESC lines [627], understanding the nuanced differences between ESCs and iPSCs proves particularly important. Much work continues to be done to probe the depth of similarity between ESCs and iPSCs [628]. Moreover, since direct reprogramming of somatic cells into iPSCs involves viral-mediated delivery of the requisite genes, efforts are under way to develop nonviral approaches to reprogramming [629, 630]. While these and a number of other basic questions remain to be answered about the nature of iPSC pluripotency and the developmental mechanisms underlying the reprogramming process, it is important to pursue avenues of inquiry involving differentiation and lineage-specific manipulation to better elucidate a role for iPSCs in cartilage tissue engineering applications. Indeed, because iPSCs can be derived from adult tissues, their use in future patient-specific therapies circumvents issues of immunogenicity associated with allogeneic cell sources and thus improves their potential for wide clinical adoption.

5.2 ASSESSMENT AND DESIGN STANDARDS FOR TISSUE ENGINEERING

A plethora of histological, ultrastructural, biochemical, and biomechanical assessments are available to evaluate engineered cartilage, with additional methods continuously being developed. Stains such as alcian blue, safranin O (often used together with fast green), and sirius red are regularly employed, along with combinations such as Movat's pentachrome and hematoxylin/eosin to histologically demonstrate the location of glycosaminoglycan, collagen, and other ECM. Distribution of specific collagens and glycosaminoglycans can be discerned using immunohistochemistry, and of great interest is whether collagen type I or II is seen as only the latter should be present in articular cartilage. Immunosorbent assays can also achieve a similar purpose by quantifying the amount of biochemical components present as opposed to showing their distribution. For implanted constructs, non-invasive methods such as MRI [631–634] and ultrasound [635–637] can be applied.

The restoration or improvement of joint function will largely result from utilizing the biomechanical properties of the engineered cartilage. This primary design standard has inspired many mechanical tests, along with mathematical models, to assist in describing both native tissue function as well the function of engineered constructs. In addition to design standards, evaluation standards are important for both researchers and for companies seeking approval from the FDA for the marketing of engineered cartilage products. As will be discussed in another section, in order to ensure safety and efficacy, the FDA evaluates scientific data generated by testing potential implants and engineered products. These data may be preclinical or clinical. Preclinical data can include specific testing protocols such as mechanical and wear testing. Certain consensus technical standards, such as those developed by ASTM International (ASTM), are recognized by the FDA, and companies can use these standard test methods in lieu of developing their own. The FDA also encourages the development and validation of computer models to be used as preclinical data in the approval process. No consensus standards exist for the evaluation of articular cartilage. Instead, this section focuses on the established procedures for evaluating the mechanical properties of the tissue, including a description of common modeling approaches used for extracting material properties from these tests. As can be seen, the models differ in both their assumptions and outputs, and it can be difficult to compare the various metrics that have been developed.

5.2.1   BIOMECHANICAL TECHNIQUES

The mechanical properties of articular cartilage can be evaluated using a variety of techniques, most of which involve monitoring the stress and strain in the tissue either over time or at different frequencies of oscillation [638]. The most common analysis techniques consider a material to be elastic, viscoelastic, or multi-phasic. These *in vitro* tests require precise experimental setups to account for boundary conditions or edge/depth-dependent effects.

Though only a few instruments are commercially available for this purpose, *in vivo* testing instruments have also been developed to evaluate cartilage stiffness. The Actaeon probe [639, 640] and the Artscan [641] are both hand-held devices that can be used arthroscopically. Using the Actaeon, the stiffness of cartilage can be measured in a fraction of a second. This instrument was verified to output data that is independent of cartilage thickness, and studies on degenerated tissue have correlated probe readings to biochemical content [639, 640]. These devices are particularly useful in combination with other monitors of a patient's health. For example, as was discussed in the

Cartilage Pathology section, changes in cartilage material properties can occur with diabetes and hormonal and steroidal levels. For patients undergoing menopause or injected with anti-inflammatory steroids, the condition of their cartilage may not be the first thing on their minds, and the availability of these hand-held devices may aide in promoting better joint health and awareness. However, most mechanical testing is performed *in vitro*, as discussed below.

For elastic measurements, the relationship between stress and strain is analyzed and then fit with a suitable mathematical model. Models are often derived that take into account the geometries of the contact region, allowing for a simpler analysis (e.g., force vs. indentation displacement data, directly obtained from the testing device).

Viscoelastic measurements are conducted using time or frequency responses. For time-based tests, the stress or strain is monitored over a set period, and mathematical models are fit to the resulting data. Stress relaxation tests observe the change in stress/force over time in response to an applied constant strain. Creep indentation tests, in contrast, observe the change in strain/deformation due to an applied constant stress. Frequency-dependent tests analyze the relationship between an applied oscillation and its signal response, thereby obtaining the storage and loss moduli, as well as the loss angle.

More complicated models of cartilage can be applied to experimental data to account for the different phases present in the tissue. These models are typically fit to data obtained from time- or frequency-dependent responses. This section focuses on the most common techniques for evaluating the compressive, tensile, shear, frictional, and fatigue properties of articular cartilage and tissue engineered cartilage constructs.

5.2.1.1   Compression Testing

A common technique for measuring the compressive properties of cartilage is through indentation [8, 642]. For this procedure, a probe of specified geometry (cylindrical, spherical, pyramidal, etc.) is indented into a material. Elastic and viscoelastic properties can be obtained using standard testing approaches (i.e., indentation and creep or stress relaxation). The indentation site should not violate any assumptions in the models. For example, models usually dictate that regions have surface characteristics allowing smooth contact and sites that are sufficiently thick.

Compressive properties can also be determined using a confined compression test [58]. In this case, the sample geometry is typically a cylindrical disk with parallel surfaces to ensure even load distributions and flush contact. The sample is tested in a confined geometry to prevent any radial expansion, thus reducing it to a one-dimensional problem. A porous platen is used to compress the sample, which allows for fluid exuded from the sample to flow through the platen-sample interface. The Young's modulus is calculated from the linear region of the equilibrium stress-strain curve, while other material parameters, such as aggregate modulus and permeability, are calculated by fitting an appropriate model to either creep or stress relaxation curves.

Unconfined compression testing follows similar procedures as with confined compression [181, 643]. However, since the sample is now free to expand radially during compression, additional parameters have to be determined to describe this two-dimensional problem. Typically, models include the Poisson's ratio as a determinant of this change. Samples are compressed using solid platens, with nothing in the radial direction except fluid, and the stress and/or strain response over time is collected.

5.2.1.2 Tensile Testing

The tensile properties of a sample can be determined from both equilibrium stress-strain measurements, as well as time-varying data (i.e., creep and stress relaxation) [644]. As with compression data, the Young's modulus can be determined from the linear region of the equilibrium stress-strain curve. The sample should be fixed firmly at the grips such that failure occurs within the working length (i.e., near the center of the sample). Specimen lengths should be significantly greater than their widths to ensure uniform strain through the working length. Extensometers or optical techniques are used to monitor the strain in the region of interest, which is plotted alongside the applied stress for analysis.

5.2.1.3 Shear Testing

Typically, shear tests are conducted on cylindrical samples in a setup similar to unconfined compression tests. A flat platen is placed on the sample, and a small tare load is applied to assure uniform contact. Shear tests can use either rotational [520, 645] or translational [70, 646] displacement strategies. As with the compressive and tensile properties, it is important to characterize both the equilibrium and dynamic responses of the sample under shear. The equilibrium shear modulus, G, is calculated from the linear region of the stress-strain curve. The dynamic complex shear modulus, G^*, is calculated using the applied and signal response to a series of oscillatory stimuli.

5.2.1.4 Friction Testing

While many theories exist describing how cartilage exhibits the frictional properties it does, most testing approaches focus on quantifying the forces present as two surfaces slide across one another. Biotribology studies of articular cartilage have focused on lubrication mechanisms at the whole joint level [647] as well as at the cartilage tissue level [648]. A variety of experimental configurations have been used, including pendulums [30, 649], oscillating arthrotripsometers [79, 650, 651], atomic force microscopy [77, 652], and plug-on-plate configurations. The latter technique is currently the most common approach. This method involves moving a sample translationally or rotationally with respect to a fixed surface or plate. Normal and frictional forces are measured, allowing calculation of the coefficient of friction. Frictional properties are sensitive to variations in bathing solution, sliding rate, and fluid pressurization within the tissue [78, 81].

5.2.1.5 Fatigue Testing

The durability of cartilage or a tissue engineered construct is perhaps the most important parameter associated with its overall functionality. Unlike the previous mechanical tests which are non-destructive, fatigue testing applies repeated loading until the sample fails [70, 653]. Usually a specific type of loading is focused on, such as compression, tension, or shear, and repeated cycles are applied until the sample is noticeably affected (cracks, fissures, tears, etc.). Fatigue life is defined as the number of cycles necessary till failure, which can depend on the applied stress, strain, and frequency.

Wear is another measure of fatigue and is defined as the removal of material from a contact surface due to mechanical effects [648]. Techniques for quantifying wear include characterizing released debris, evaluating surface topography, and imaging the bulk tissue. The severity of a damaging abrasion can be determined by measuring the size of released debris as well as the depth of penetration at the surface [654]. Cartilage roughness, as determined using a variety of scanning microscope techniques, can indicate how well the material will perform under shear or friction. Other imaging techniques that look at the tissue as a whole can be used to evaluate not only the surface characteristics but also any breakdown of the tissue below the surface.

5.2.1.6   Mathematical Models of Articular Cartilage

Mathematical models are used to interpret results obtained from carefully designed evaluation tests, such as those described in the previous section. By fitting a model to experimental data, a quantification of the mechanical properties can be achieved. Numerical representations of mechanical characteristics are of critical importance for comparison among studies, and researchers typically use similar testing techniques to facilitate this. Properties such as the Young's modulus, coefficient of friction, and streaming potential are just a few of the characteristics that can be used to describe the natural function of articular cartilage.

Some mathematical models are very basic in their description of cartilage while others are extremely complex. It is important to remember, however, that they are all only representations of how the tissue might function and do not replicate every possible intricacy. Even simple models can provide valuable information, though, and can serve a purpose in evaluating a subset of properties. For example, the elastic components of a material can be described by:

$$E = \frac{\sigma}{\varepsilon}$$

where E is the Young's modulus, σ is the stress, and ε is the strain. Modeling cartilage just as an elastic material can provide a measure of its elastic response, but it might not correspond as accurately with experimental data as more complex models. Combinations of elastic and viscous elements can help to describe a material with time- or frequency-dependent responses. The viscous components can be modeled by:

$$\eta = \frac{\sigma}{\frac{d\varepsilon}{dt}}$$

where η is the viscosity coefficient and $d\varepsilon/dt$ is the time derivative of strain. By using elastic and viscous elements, viscoelastic models can be derived.

Biological materials are typically considered to be viscoelastic since their deformation characteristics vary with respect to time and/or frequency. While perhaps also not technically appropriate, articular cartilage can be modeled as a viscoelastic material. By fitting to either stress relaxation or creep data, parameters can be extracted that describe the time-dependent response of a material. Simple models of viscoelasticity include Maxwell (spring-dashpot in series) and Kelvin-Voigt (spring-dashpot in parallel). An extension of these models is the Kelvin model, or standard linear solid (spring in parallel with a spring-dashpot), whose deformation response during a creep/relaxation test is represented by:

$$\left(1+\tau_\varepsilon \frac{d}{dt}\right)\sigma = E_R\left(1+\tau_\sigma\frac{d}{dt}\right)\varepsilon$$

where E_R is the relaxed modulus, τ_ε is the relaxation time for constant strain, and τ_σ is the relaxation time for constant stress. The spring elements in the model describe the stiffness of the tissue while the dashpot helps describe the time-dependent deformation. When fit to experimental data, these can be used to calculate the instantaneous modulus and apparent viscosity of the material:

$$E_0 = E_R\left(1+\frac{\tau_\sigma-\tau_\varepsilon}{\tau_\varepsilon}\right)$$

$$\mu = E_R\left(\tau_\sigma-\tau_\varepsilon\right).$$

While viscoelastic models are useful for providing a basic description of biological tissue deformation, they are not particularly representative of the actual mechanical characteristics associated with articular cartilage. As discussed previously, cartilage can be described as having two phases: one solid, one fluid. More complex mathematical models of the tissue, such as the biphasic/poroelastic model [58, 655], take into account this composition, providing parameters that describe the stiffness, fluid flow, and deformation characteristics of the tissue. One example of this is the following equation that describes confined creep compression of cartilage using the biphasic solution [58]:

$$\varepsilon_{zz}(t) = \frac{F_0}{H_A}\left[1-\frac{2}{\pi^2}\sum_{n=0}^{\infty}\left(n+\frac{1}{2}\right)^{-2}\exp\left\{-\pi^2\left(n+\frac{1}{2}\right)^2 H_A kt/\left[(1+2\alpha_0)h^2\right]\right\}\right]$$

where ε_{zz} is the observed strain, F_0 is the applied constant load, H_A is the aggregate modulus, k is the permeability, h is the sample thickness, and α_0 is the solid content ratio. When fit to experimental data, values for H_A and k can be extracted. Other equations exist in the literature for the various geometries and device configurations that are possible for cartilage testing.

More complex models of articular cartilage exist and are useful for identifying particular parameters that might be of interest. For example, an alternative to the biphasic model of cartilage is the poroviscoelastic model, which accounts for the different phases of the tissue and well as their short and long time responses to loading [46]. Articular cartilage can be modeled in even more complexity than as just a two-phased tissue. A third phase, the ionic phase, can significantly affect the motion of fluid through the solid matrix, and hence, the deformation characteristics of the tissue. The triphasic model accounts for contributions from the solid, fluid, and ionic phases of the tissue but results in the same parameters as the biphasic solution, with the addition of fixed charge density [656].

Other characteristics that might be of interest include the frictional and torsional properties of the tissue. The coefficient of friction, μ, can be determined using a simple relationship between the normal, N, and friction, F_f, forces measured during friction tests:

$$\mu = \frac{F_f}{N}$$

Friction tests are applied by sliding a probe across the cartilage surface and collecting data for the two forces of interest. This can be done at the macroscale level of the joint [657] or the microscale [77] to determine the frictional characteristics of the sample. An alternative approach is to measure friction using a rheometer, which moves two surfaces rotationally. This device can also measure both the simple and dynamic shear properties of cartilage. Simple shear testing can provide a measure of the shear modulus, G:

$$\tau = G\gamma$$

where τ is the shear stress and γ is the shear strain. Oscillatory measurements are used to determine the storage and loss moduli, as well as the complex shear modulus, G^*, which is simply the sum of the two.

5.2.2   DESIGN STANDARDS - FUNCTIONAL IMPROVEMENT VERSUS REGENERATION

The determination of design standards for articular cartilage tissue engineering largely depends on the type of restoration sought after. It appears obvious that long-term, fully functional restoration is desirable, but whether this is to be achieved immediately or some amount of time post-implantation will alter the design standards. Using non-terminally differentiated chondrocytes will also result in additional considerations. Nonetheless, of prominence are (1) mechanical properties, (2) biochemical properties, (3) integration, (4) construct size, (5) contour, and (6) ease of implantation.

The biomechanical properties of native tissue has long been well-characterized in different anatomical locations, at various ages, diseased states, and under various hormonal and drug concentrations to provide ample direction for tissue engineered constructs to emulate. As it has been identified that cartilages of different anatomical locations can have different properties [50–52], should a cartilage product be tailored specifically for the talus and another for the knee, and at what point does stiffness mismatch result in ill-borne stresses that can lead to articular damage? If the engineered cartilage is softer than native tissue, can functional improvement nonetheless be achieved, or will the construct break down like the mechanically inferior fibrocartilage? These are all questions that should be considered.

There are other issues that complicate articular cartilage regeneration when using implanted grafts or cells. Integration between the implanted construct and the surrounding tissue is of crucial importance for mechanical function since the cartilage surface acts as one entity to distribute applied loads. The interface between new and old tissues is often weak, especially on the surface, and failure is probable unless sufficient healing takes place that helps to integrate the two tissues. Ideally, implanted constructs should account for the tissue microstructure and create a replacement that has correctly aligned collagen fibers and regional variations [658]. The alternative is complete remodeling of the tissue *in vivo*, which may never occur in an environment that has only a limited repair capacity. Various studies have been performed to examine the integration of cartilage to cartilage or tissue-engineered constructs to cartilage, and the general consensus regarding the main factors that hinder integration are the following:

- Cell death at the wound edge, even in surgically prepared defects, results in metabolically inactive tissue with antiadhesive properties that prevent cell adhesion and migration to the injury site [659–663].

- Insufficient numbers of viable cells do not synthesize integrative matrix between the two surfaces to be joined [661–666].

- Cell migration across metabolically inactive tissue is hindered by dense collagen [659–663].

- Insufficient construct and tissue collagen crosslinks do not allow integration [457, 667, 668].

Proposed solutions to these obstacles include implanting less mature (and, therefore, softer) constructs that contain a large number of cells at the construct edge, using enzymes to partially degrade the collagen [659, 661, 669–671] and glycosaminoglycans [659, 661, 662, 664, 672–674] of the tissue, and applying agents with the aim of promoting native-to-transplant collagen crosslinking [457, 667, 668, 675–677].

With the generation of a sufficiently large construct comes the requirement of a sufficient number of cells or time in culture to expand them. As discussed in the previous section, research is currently being conducted with MSCs, ADSCs, ESCs, iPSCs, and DIAS to examine their suitability for cartilage resurfacing. Whereas the design standard for construct size is simple (the construct needs to be thick enough or plentiful enough to fill the lesion volume), desirable characteristics for cell sources are more complex. It is desirable for the cell source to have a short time in culture, be cost effective to expand, be homogeneously differentiated, and be effective in producing functional constructs. Lastly, the proper geometry and contour, ease of implantation, shelf-life, and the pathway to regulatory approval and to the market are additional issues to be considered in designing a viable process for articular cartilage repair.

5.3   CURRENT AND EMERGING THERAPIES

In this section, current therapies that involve biological products for cartilage repair are described. It is important to note that treatment may include the cartilage, bone, synovium, and muscles, as it is unclear at this time whether any one component is the primary cause for OA progression [678]. As the last treatment option, total joint replacement, or arthroplasty, will only be briefly mentioned here. Improvements in the design of arthroplasty implants have significantly lengthened their usable lifespan, though failure can still occur due to a variety of reasons. Aside from the wear of the articular surface and the failure of the implant itself, failure can result from the mechanical differences between the implant and the surrounding bone. As the stiffer implant is capable of bearing more load, shielding of the surrounding bone occurs, and osteolysis, instability, and implant loosening can occur. The difference in stiffness between the artificial and natural materials can also result in periprosthetic fracture. As outlined in Chapter 2, children and adolescents are not good candidates for these procedures, both due to their developing skeletons and to the fact that their age requires a longer solution. In addition, arthroplasty is not the best solution for cases where focal lesions are concerned, as often caused by sports injuries. Thus, this section discusses non-arthroplasty therapies, divided into non-surgical and surgical techniques. Surgical transplantation of allogeneic or xenogeneic materials is discussed in the next section with cartilage immunology, as the outcome of these transplants relies heavily on whether the implant is accepted by the body.

5.3.1   NON-SURGICAL METHODS

Articular cartilage injuries can be caused by a variety of reasons, as reviewed in Chapter 2. Since sports or improper loading are implicated, activity modification, weight and/or body fat loss, physical therapy, or the use of a cane to lessen the load applied on the problematic joint are all non-surgical methods in addressing discomfort. For instance, weight reduction in combination with strength exercises have been shown to significantly reduce knee pain in overweight and obese individuals [679]. The reduction of body fat independent of weight loss has also been shown to be beneficial [680,681]. Injections include visco-supplementation (e.g., hyaluronan) and corticosteroids (though it has been shown that some injections can soften the remaining cartilage [162]). Aspirin, ibuprofen, and a variety of COX-2 inhibitors are medications that are also often used, as well as dietary supplements. It has long been believed by many that items such as copper bracelets can alleviate joint pain caused by arthritis. Whereas it has already been determined that the placebo effect is at work in this case [682], the efficacy of other devices, such as magnetic bracelets [683], is still undergoing investigation.

The popularity of copper and magnetic bracelets stems from several desirable characteristics, namely, the ease of use, non-invasiveness, the presence of few or no side effects, and the perception that these items are more "natural" than other therapies. Over-the-counter medication such as aspirin and ibuprofen share many of these desirable characteristics, as do dietary supplements such as glucosamine and chondroitin. The latter two are of particular interest due to their recent surge in popularity. These products have been recommended by both OsteoArthritis Research Society International in 2007 [684] and the European League Against Rheumatism in 2003 [685] while being the subject of conflicting results and heated debates. Questions have been raised about study results with regard to design, sample sizes, publication bias, and the choices of controls, all complicated by the overwhelming variety of formulations and derivatives available on the market. The Glucosamine/Chondroitin Arthritis Intervention Trial (GAIT) [686], funded by the National Institutes of Health and published in 2008, showed that these supplements did not significantly reduce pain as compared with a placebo, though this study has itself received similar criticism as listed above. Efficacy is still under investigation, and these supplements have shown benefits in a recent meta-analysis [687].

5.3.2   SURGICAL METHODS

Though it may be ideal to not have to compromise the joint by introducing foreign matter, surgical methods also provide ways to alleviate pain and to restore function.

5.3.2.1 Debridement

Arthroscopic abrasion arthroplasty is a procedure where the cartilage defect is smoothed and re-shaped. Burrs, diseased tissue, delaminated cartilage, and flaps can be removed to improve the gliding motion and to provide temporary relief. Osteochondral defects can also be treated by removing the dead bone or sclerotic lesions to result in fibrocartilage production. The fibrocartilage repair tissue has been reported to last up to six years [688]. Reports have shown that, for chondromalacia patellae, 75% of the patients were satisfied with the procedure when followed up [689]. Even for athletes, the procedure has shown quick success, with resumption of sports activities on an average of 10.8 weeks following the procedure [690]. Cartilage debridement has been applied to various joints, including the knee [691, 692], elbow [693, 694], ankle [695], and shoulder [696], oftentimes in combination with other procedures.

5.3.2.2 Microfracture

As articular cartilage possesses little intrinsic healing response, the introduction of mesenchymal stem cells and growth factors via microfracture from the subchondral bone has been widely employed for isolated chondral defects of the knee [697, 698], shoulder [699], and ankle [700]. The damaged cartilage is first removed down to the calcified zone to expose healthy adjacent tissue. The calcified cartilage is then removed, and evenly spaced microfractures into the subchondral bone are introduced. Blood then fills the defect, resulting in a fibrin clot that initiates a healing response as described in Chapter 2 The repair tissue is thus fibrocartilaginous in nature, with inferior material properties as compared to healthy articular cartilage. Insufficient repair tissue (i.e., too thin) can also result in altered biomechanics that lead to degeneration. To improve upon this, microfracture has been used in combination with coverings, such as a periosteal flap [699] or natural (chitosan [701] and collagen scaffolds [702, 703]) and synthetic materials (e.g., PGA and hyaluronan [704]). Other improvements to the technique include the addition of a BMP-4 carrier, which showed more rapid repair [705]. BMP-7 has been shown to increase the volume of repair tissue generated [706]. A recent systematic review of 28 studies describing 3,122 patients has shown that the procedure is effective within the first 24 months in improving knee function [707]. Subsequently, the effectiveness wears off, especially for patients forty years and older [707, 708], as the fibrocartilage formed can eventually degenerate, resulting in recurred loss of function.

5.3.2.3 Autologous Implants

Though limited in source, autologous implants enjoy several advantages as transplant materials such as not eliciting immune responses and having functionality close to the tissues they are replacing. Taken from non-load bearing regions, autologous implants may be less stiff but contain live, autologous cells to potentially allow for continued remodeling. Unlike allogeneic tissue, the concern for disease transmission is greatly mitigated in this case. However, the scarcity of source material, donor site morbidity, differences in shape between the implant and the recipient site, and, significantly, the need for multiple invasive surgeries (first to retrieve the implant, then to insert it), limit the use of autologous implants. Chondral implants face the significant problem of integration, and autologous implants typically fall into two forms, osteochondral plugs and autologous cells.

Osteochondral plugs are harvested from non-weight bearing regions and have shown efficacy for as long as ten years [709]. To better fit the differences in curvature between the donor and recipient surfaces, mosaicplasty [710] has been shown to have better results than Pridie drilling, abrasion arthroplasty, microfracture repair, and autologous chondrocytes implantation [710–712], as evaluated by radiography, MRI, biopsy analysis, and other techniques to generate scores from a variety of systems, including the International Cartilage Repair Society scoring system. Good to excellent scores have been demonstrated for resurfacing the femoral condyle (92%), tibia (87 %), patella and/or trochlea (79%), and talus (94%) using mosaicplasty [713].

Attempts have been made to use less tissue to reduce donor site morbidity and to increase the amount of usable grafts. For example, 6 mm diameter osteochondral plugs have been placed into 10 mm diameter defects in the sheep model to examine whether the resulting 2 mm band around the plug can be filled with repair tissue [714]. Tissue ingrowth in this ring was observed, but it consisted of fibrocartilage. Since a 6 mm diameter is below the 7 mm "critical defect" size [215], a donor site defect of this size or below is expected to recover. However, it has been shown that the repair tissues at the donor sites consist of fibrocartilage, and approximately 3% of the sites exhibit morbidity [713].

Though osteochondral plugs are taken from non-weight bearing locations, the material properties of adjacent repair fibrocartilage can still nonetheless be inadequate for long-term use. For instance, fibrocartilage repair tissue (and osteoarthritic tissue) has been shown to lack the extent of collagen organization and alignment [715]. Even under low loading conditions, the shear that is an integral aspect of articulation still may prove challenging to the relatively unorganized repair tissue. Without proper organization and material properties, the repair tissue from the donor site can degenerate. For this reason, osteochondral autografts are typically employed after other techniques have been excluded due to their complexity or inadequacy.

Currently, the only FDA-approved autologous cultured chondrocyte product is Carticel®, developed by Genzyme Corporation [716,717]. Cartilage is first harvested from low-weight bearing regions of the knee and sent to Genzyme for enzymatic digestion. The released chondrocytes are expanded *in vitro* for several weeks and sent back to the surgeon. Prior to the implantation of these cells, the defect must first be debrided and cleaned, and a periosteal flap must also be harvested. This flap is stitched over the defect to form a pocket, into which the expanded cells are injected [718]. The tissue that forms from these cells can be hyaline or fibrocartilage, with as little as 15% of the cases reporting hyaline cartilage [719, 720]. Collagen type II that does form in this case is not aligned like native tissue [721]. As a result, the stiffness of the repair tissue has ranged from 62% [722] to 90% [723] of the values of the surrounding cartilage.

Autologous cell implantation (ACI) has been shown to yield better results for femoral cartilage defects than patella or tibial defects, with the location within each region also affecting clinical outcomes [724, 725]. It has been shown that, four years after ACI, 75% of patients were mostly satisfied with the surgical outcome [726]. When comparing to microfracture, conflicting studies exist.

Better clinical results have been seen for ACI than microfracture although both show satisfactory outcomes at medium-term follow-ups [727].

Clinical studies involving joint repair have long established that immobilization inhibits long-term healing in articular cartilage [728–731]. Motion is needed to induce movement of fluid and nutrients throughout the joint spaces, as well as providing mechanical cues that can stimulate the chondrocytic phenotype. However, the rigorous mechanical environment of the joint is too challenging for newly formed tissues, often resulting in rapid failure of the implanted constructs. Continuous passive motion is commonly used for the first two weeks after surgery to facilitate the transport of fluid, nutrients, and solutes within the joint, thereby stimulating chondrocyte metabolism [731]. Passive motion alone may be insufficient for cartilage healing, though, since it does not allow any significant loading of the tissue [730]. Active motion, including incremental strength and weight bearing exercises, may be necessary to stimulate repair processes during rehabilitation. As in many *in vitro* experiments, chondrocytes respond best when suitable mechanical forces are present.

As discussed previously, cartilage resurfacing alone may be insufficient in completely addressing lesions as other problems may underlie the lesions' formation. The same can be said of therapies using either mosaicplasty or ACI. As joint malalignment can result in lesions, the post-operative joint alignment can affect ACI outcomes [732]. Other factors include concomitant treatments, such as different rehabilitation regimens [733, 734], and, of course, patient age [735] and surgical history [736].

5.3.2.4 Osteotomy

The surgical procedures described thus far are oftentimes required due to lesions that result from improper mechanics. In the cases of malalignment, contact pressures of the defect area may be reduced to physiological levels by osteotomy. In this case, methods include anteromedial transfer of the tibial tubercle to decrease the contact forces on the lateral facet of the patella [737], as anteromedialization of the tibial tubercle has been shown to lower contact pressures in the lateral trochlea in a cadaver model [738]. This, in combination with other surgical techniques such as ACI, can be beneficial to patients with multiple knee disorders [739]. Currently, the combined consideration of alignment with autologous osteochondral grafting are associated with traumatic reconstruction case studies and are thus difficult to compare [740]. Nonetheless, in order to prevent relapse, future cartilage resurfacing methods may consider the accompaniment of additional orthopaedic adjustments.

5.3.2.5 Other Treatments and Emerging Techniques

An alternative to *in vitro* tissue engineering is to implant a scaffold *in vivo*, with or without cells, and allow regeneration to occur with minimal additional manipulation. Regeneration of hyaline cartilage within the body is complicated by the rigorous mechanical environment present in active joints. However, researchers are currently investigating means to accomplish this goal since overall healing time is anticipated to be less for *in vivo* versus *in vitro* tissue engineering approaches. Furthermore, the complex mixture of biochemical and biomechanical cues present in the body can accelerate tissue growth that is difficult to produce otherwise. Most of the factors previously discussed for *in vitro* tissue engineering are applicable to *in vivo* tissue engineering, except for some types of mechanical stimulation. However, loading is applied naturally by the normal physiological environment, which could be termed the ideal cartilage bioreactor.

The *in vivo* growth of cartilage tissue depends on many different factors that cannot be modeled well *in vitro*. For example, many cytokines and bioactive molecules exist in living bodies that cannot be easily included in a laboratory experiment, either because they are still unknown, or more practically, their sheer numbers are unreasonable for controlled studies. Using serum in the culture media is intended to replicate these conditions somewhat, but results can be dramatically different when the growth environment is an active, living body which can endogenously produce bioactive molecules in response to the implanted construct. Even in an isolated tissue such as cartilage, chemical and mechanical signals can impact the development of the tissue whether it is an empty defect or an implanted construct.

Researchers investigating *in vivo* cartilage engineering have focused on repairing defect sites with transplanted cartilage/cells, synthetic materials, and cell-seeded cross-linkable scaffolds. The first includes autologous and allogeneic cartilage/cell implantation, in which cells or minced tissue are inserted into a defect site and then kept in place with a covering, such as a periosteal flap [718]. The second approach is primarily a stop-gap measure that would provide a mechanically functional insert but does not allow regeneration of the tissue [741–743]. The third repair technique includes several different types of synthesized polymers that can transition from a fluid to a stiff gel using either light or heat as an initiator [417, 744]. The mentioned *in vivo* repair techniques all have advantages and disadvantages although none have resulted in long-term, functional repair of articular cartilage that is comparable to healthy tissue.

ACI has been modified in various ways in animal studies. Instead of a cell slurry, expanded chondrocytes have been condensed into spheroids first and then implanted into SCID mice [745]. Attempts have also been made by embedding cells in an alginate-gelatin hydrogel with subsequent implantation in sheep. Hyaline-like repair tissue formed in both cases although better histological scores resulted when chondrocytes were included [746]. Attempts have also been made at replacing the periosteum flap with other materials, such as collagen sheets with embedded cells [747, 748], and it has been shown that symptomatic hypertrophy, disturbed fusion, delamination, and graft failure observed with periosteum use can be subsequently reduced [749].

A possible approach to *in vivo* articular cartilage replacement is to insert synthetic constructs, which would fill a defect and provide mechanical support and a low-friction surface. These cell-less constructs would be non-resorbable and could likely find a niche as a stop-gap measure for patients wanting to delay full arthroscopy procedures [742]. While not technically considered tissue engineering, synthetic replacements do represent an attractive option due to their ease of handling and modification. Synthetic replacements can provide structural support for limited periods of time, but eventually more drastic procedures will be necessary as the conditions in the joint continue to degrade. A scaffold that would not necessarily bear weight initially but would instead offer ease in implantation would be photopolymerizable hydrogels. These allow for minimally invasive implantations of the cell/polymer constructs [744].

Several cell-scaffold combinations have been examined for resurfacing. BioSeed-C, a fibrin and polymer-based scaffold (PGLA/polydioxanone), has been able to mitigate pain and improve knee-related quality of life measures after one-half, one, and four years [750]. A collagen type I/III mesh with chondrocyte implantation has shown good or excellent outcomes two years post-operation in 82% of patients that underwent the procedure although 75% of the defects showed fibrous tissue formation rather than hyaline [751]. Hyaff-11, by Fidia Advanced Biopolymers, is an esterified hyaluronic acid scaffold that could be implanted using an arthroscopic technique to yield formation of hyaline-like cartilage tissue [752]. Outcomes for these products and procedures can depend on the patient population. For instance, young, highly-active patients had better outcomes than less-active counterparts after treatment with Hyalograft C [753,754].

Future work could take advantage of *in vivo* tissue engineered constructs using the scaffolds described above. Alternatively, the scaffoldless self-assembly of chondrocytes may be employed. In this case, researchers rely only on cells and the biological/mechanical signals that are necessary to induce a chondrogenic response. Foreign scaffold materials would be unnecessary, and one of the advantages of *in vitro* tissue engineering is that a construct with sufficient mechanical properties would be delivered using such an approach to withstand physiological loading.

5.4   IMMUNE RESPONSE, IMMUNOGENICITY, TRANSPLANTS

Being alymphatic and avascular, the tough, hyaline matrix of articular cartilage prevents easy access to cells embedded within, regardless of the origin of these cells. Taking advantage of these characteristics, proposals have been made to use cartilage matrix as a barrier to protect transplanted cells [755]. For the same reason, the joint is considered by some to be "immune-privileged" due to the body's limited ability to detect and reject implanted tissue [756]. However, cartilage matrix is not itself without rejection issues [756,757]. Collagens type II, IX, and XI and proteoglycan core proteins all have antigenic properties [758–762]. The chondrocytes, too, have been found to contain major histocompatibility complex (MHC) Class II antigens, which can elicit a cell-mediated immune response as described below [763,764]. Natural killer cells can also attack chondrocytes [765–767].

This section deals with additional therapies and possibilities that follow from the previous section. Future clinically applicable treatments may not deal solely with this tissue but the subchondral bone as well. The rationale for cartilage immuno-privilege will be discussed, and studies involving cartilage or osteochondral transplantation will also be presented. The difference is that, in this case, the materials are of a foreign nature whether allogeneic or xenogeneic. These include both cells and tissues transplanted into the joint, and to understand the body's reaction, we must first have a basic understanding of the rejection process. First, the mechanism of rejection will be presented in stages. From this perspective, the reactions to the introduced cartilage grafts will be discussed.

5.4.1   CELLULAR AND HUMORAL RESPONSES

Both cellular and humoral responses can be directed against implanted cartilage from genetically-disparate donors. For a cellular response, which is mediated by T lymphocytes, a sensitization phase occurs when an antigen (ex. Virus or implant) is recognized by a macrophage and presented to the T helper 1 cells (both CD4+ and CD8+). The transplant may contain antigen presenting cells (APCs) that express appropriate antigenic ligands on their MHC receptors, and this, in addition to a required co-stimulator signal, activates the T cells. The immune system is also stimulated as nonlymphoid passenger leukocytes migrate from the graft tissue to the lymphoid organs. Cytokines then initiate the proliferation phase, where cytotoxic T cells multiply against the antigen. Passenger leukocytes undergo maturation from immature dendritic cells to mature APCs that activate an array of T lymphocytes, including CD4+, CD8+, and naïve T cells during migration. Effector immune responses then proceed to defeat the antigen. Activated T cells secrete various cytokines (i.e., IL-2, IFN-γ, TNF-β, etc.) to recruit a variety of other host immune cells, inducing increased expression of MHC Class I and Class II molecules by donor cells. For instance, the autocrine response of IL-2 results in cytotoxic T lymphothocytes that attack the APCs; macrophages are recruited to the graft site by IFN-γ, and TNF-β has a direct, cytotoxic effect on graft cells. With the antigen defeated, the response is then downregulated by T suppressor cells, and memory T cells mature for future recurrences [768–771].

The humoral response is generally directed against bacteria, though implant rejection can also occur via this response and is mediated by B lymphocytes. The naïve B cell recognizes the bacteria and presents it to T helper 2 cells. Cytokines then induce B cells to produce antibodies, which can work in several ways in neutralizing the antigen. With the antigen defeated, the response is then downregulated, and memory B cells form. The antibody can prevent the bacteria from adhering by surrounding it. Antibodies can also promote opsonization, whereby the antibody promotes phagocytosis. Compliment, which enhances opsonization and can lyse some bacteria, may also be activated by antibodies [770].

5.4.2   ALLOGENEIC TRANSPLANTS

As mentioned previously, techniques exist that use autologous chondrocytes and osteochondral plugs, and two of the major limitations of these techniques are repeated surgeries and limited donor tissue. Allogeneic sources have thus been considered to circumvent these two issues. The use of allogeneic tissue does present an elevated risk of disease transmission, and testing for diseases such as HIV, hepatitis, and syphilis must first be performed. Immune responses against the cells and tissues are also problematic.

Allogeneic chondrocyte transplantation has proven to be difficult due to the humoral response mounted against these cells [772]. Allogeneic chondrocytes implanted into posterior tibial muscles formed nodules that immediately attracted macrophages, and natural killer and cytotoxic/suppressor T cells were also recruited to destroy the nascent cartilage over time [763]. This slow destruction of the repair tissue has been shown in several other studies [773–776]. Cells embedded in matrix, however, were relatively safe. Implantation of chondrocytes versus chondrocytes allowed to first produce matrix has shown that the immune response was greatly diminished in the latter case [777].

In animal models, when allogeneic chondrocytes are implanted with their ECM into rat, rabbits, or dogs, neither significant leukocyte migration nor cytotoxic humoral antibodies were observed for several studies [772,778,779] while others have shown increased presence of inflammatory mononuclear cells and less repair cartilage in defects filled with antigen-mismatched transplants in dogs [780]. Due to the potential for immune responses, and due to the lack of availability of fresh cadaveric donor tissue, frozen or pressure washed osteochondral allografts have been examined. In these cases, the cells are likely dead, thus reducing the immunogenic response [781,782], but the grafts are biochemically and histologically inferior to fresh grafts. Cryopreservation does allow for tissues to be banked, giving greater time for screening of diseases in the tissue [783,784].

Allografts have shown considerable success; long-term follow-ups of up to fifteen years of patients receiving fresh osteochondral allografts revealed allograft survival rates of 75%-95% at five years, 64%-80% at ten years, and greater than 60% during fourteen and fifteen years [785,786]. However, compared to unipolar repairs, clinical trials have demonstrated that allograft implantation is unsuitable for bipolar lesion repairs, with 50% of grafts failing at six years (as compared to 84% in unipolar repairs) [787,788]. Cryopreserved and frozen allografts have yielded good to excellent scores following transplantation in roughly 70% of patients up to four years. It is worth noting that, while success has been demonstrated in the treatment of condylar lesions using allografts, the procedure is still considered to be a salvage operation and is currently only suited for young, active patients with isolated patellofemoral articular cartilage disease, for whom previous procedures have failed.

Though currently not available as therapies, emerging technologies employing *in vitro* tissue engineering has shown much success when allogeneic cells are combined with scaffolds. Implantation of allogeneic chondrocytes embedded in collagen [789–791], agarose [792], and PGA [352,793], among others, have been examined in various animal models. In general, the hyaline histological appearances were found with little to no sign of immunologic reactions. Similarly, the implantation of allogeneic MSCs in a hyaluronic acid-based gel in a caprine model has shown only mild immunologic rejection [794]. As seen with the contrast between chondrocytes alone and chondrocytes with associated matrix, *in vitro* seeding and culture of these tissue engineered constructs allow for the formation of a protective, hyaline-like matrix around the cells prior to implantation, boding well for the future of tissue engineering therapies utilizing allogeneic cells.

5.4.3   XENOGENEIC TRANSPLANTS

While allogeneic tissues are more easily procured than autologous tissues, xenogeneic tissues are of even greater abundance. In this case, the source is of a different species, and immunological concerns are further heightened. No cartilage product using live xenogenic cells currently exists though the methodology is being examined in several animal models. Rat chondrocytes implanted in rabbit muscle resulted in the complete destruction of the implant by macrophages and giant foreign body cells [795]. However, since cartilage is immune privileged, xeno-implantation into an articular defect yields better though sometimes mixed results. Using fibrin glue as a matrix, rabbit chondrocytes transplanted into goats resulted in mild synovitis and the formation of fibrous repair tissue [297]. Implantation of pig chondrocytes into osteochondral defects of adult rabbits resulted in the production of hyaline-like tissue with the absence of inflammatory cells [796]. The immunogenicity of chondrocytes from transgenic pigs has also been examined *in vitro*. Chondrocytes isolated from H-transferase transgenic pigs have been shown to have lower expression of the Galalpha1,3Gal antigen (alphaGal) that humans reject [797], and as a result, experience lowered compliment deposition and monoblast adhesion [798]. Aside from chondrocytes, no immune reaction was found when human MSCs were implanted into a swine model to restore the articular surface [799]. This may be in part because MSCs have been shown to display immunosupression properties when combined with IFN$_\gamma$, TNF, IL-1$_\alpha$, or IL-1$_\beta$ via nitric oxide production [800].

In addition to cells, xenogeneic tissues have also been examined. It is believed that decellularizing xenogenic tissue will be a viable option for the generation of replacement tissue as the antigenic intracellular proteins and nucleic acids are eliminated while preserving the functional properties of the tissue's extracellular matrix [801]. Ideally, the biomechanical properties of the tissue will also be preserved. For instance, an acellular dermal matrix [802] has seen successful clinical use as the FDA-approved Alloderm product or the porcine derived Strattice product. Additionally, acellular xenogenic tissues have been created for many musculoskeletal applications, including replacements for the knee meniscus [803], temporomandibular joint disc [804], tendon [805], and ACL [806], as well as in other tissues including heart valves [807–813], bladder [814], artery [815], and small intestinal submucosa [816, 817]. Previously, a photo-oxidation approach was used for bovine xenograft decellularization, followed by implantation *in vivo* into a sheep model [818]. The photo-oxidation approach, which resulted in nonviable chondrocytes without elimination of DNA or antigen reduction, resulted in a reduced monocyte and plasma cell infiltration in the implant after 6 months. Various other chemical treatments have been developed, such as 1% sodium dodecyl sulfate (SDS), 2% SDS, 2% tributyl phosphate (TnBP), 2% Triton X-100, and hypotonic followed by hypertonic solution [803–813, 816, 817]. These methods were applied to self-assembled tissue-engineered cartilage constructs and cartilage explants [819, 820]. All SDS treatments resulted in cell removal histologically, but 2% SDS for 1 h decreased DNA content by 33% while maintaining biochemical and biomechanical properties. Additionally, 2% SDS for 8 h resulted in complete histological decellularization and a 46% reduction in DNA content, although compressive stiffness and glycosaminoglycan (GAG) content were significantly compromised. As these are the only investigations of decellularized cartilage, clearly there is a dearth of studies demonstrating the effects of tissue decellularization on cartilage as well as engineered cartilage constructs.

In a decellularization process, epitope removal must be considered in addition to removal of cell and nuclear materials. In general, the implantation of xenograft tissue results in hyperacute rejection of the implant, in which pre-existing host antibodies bind to endothelial cells, leading to vascular collapse of the xenograft [821, 822]. Typically, joint tissues such as cartilage are not subject to hyperacute rejection due to the avascularity of the tissue; thus, cartilage is considered to be relatively non-immunogenic [823, 824] . However, the cartilage matrix, particularly the α-galactosyl carbohydrate structure, or epitope, is a source of immunogenicity that leads to tissue destruction with xenograft implantation, but only in humans and primates [825–828]. For instance, implanted porcine and bovine articular cartilage in cynomolgus monkeys elicited extensive humoral response to the xenografts, leading to chronic graft rejection with fibrous encapsulation and peripheral leukocyte infiltration [829]. In a follow-up study, implanted porcine or bovine articular cartilage in cynomolgus monkeys resulted in the increase of anti-α-galactosyl IgG by up to 100-fold, accompanied by increased complement-mediated cytotoxicity; thus indicating a chronic rejection response to the tissue [830]. Finally, porcine articular cartilage pre-treated with α-galactosidase to remove the α-galactosyl epitope resulted in a significant reduction in the inflammatory response to the xenograft and decreased T lymphocyte infiltration into the tissue in the cynomolgus monkey [831].

As only humans and primates do not express the α-galactosyl epitope and, therefore, produce anti-α-galactosyl antibodies, an α-1,3-galactosyltransferase knockout mouse has been developed as a small animal model to study the immune response to this epitope [832]. These mice mimic humans and primates in that they do not produce the α-galactosyl epitope as the α-galactosyl epitope is formed by α-1,3-galactosyltransferase. This model has been used to assess *in vivo* immunogenicity and response to α-galactosyl production of a decellularized vascular graft [833] and may be employed to examine xenogeneic cartilage in the future.

5.5   BUSINESS ASPECTS AND REGULATORY AFFAIRS IN CARTILAGE TISSUE ENGINEERING

A bulk of this text has thus far been concerned with how tissue engineered constructs can be produced. Growth factors and bioreactors, used in combination with a variety of cells and scaffolds, are intended to result in functional cartilage implants. Current therapies, including autologous, allogeneic, and xenogeneic transplants have been presented, as well as other emerging technologies that do not employ cells or tissues at all. Depending on the indication, successful articular resurfacing may eventually be achieved using a variety of products and methods. These products might also include glues and fixatives specifically developed to adhere a piece of engineered cartilage to the native tissue, as well as tools and enzymes that prepare the osteochondral or chondral defect to receive an implant. The above are but a short, categorical list of the treatment possibilities that can come to market. Most of the cartilage tissue engineering technologies described in this text are still in their experimental stages, and their safety and efficacy are yet to be verified. The pathway potential products can take in demonstrating safety and efficacy will depend on the nature of the product and its intended use. This is an important consideration from a business perspective because some routes take longer and cost more than others. A brief introduction to the regulatory bodies of the US Food and Drug Administration (FDA) will be presented, and pathways to regulatory approval will be provided in this section.

5.5.1   REGULATORY BODIES

In the US, the FDA is charged with protecting the public health by assuring the safety, efficacy, and security of drugs, biological products, medical devices, radiological products, cosmetics, and domestically produced and imported foods. There are seven product-oriented Centers within the FDA to evaluate different classes of products, such as drugs, vaccines, or medical devices. A product that is a combination of more than one class, e.g., a medical device that releases drugs, is assigned by The Office of Combination Products to one of the seven Centers where primary jurisdiction over the product will reside. For cartilage therapies, four out of these seven Centers are typically involved.

Through the regulation of food, cosmetics, and dietary supplements, the Center for Food Safety and Applied Nutrition (CFSAN) protects the nation's health by ensuring that these products are safe, sanitary, wholesome, and honestly labeled. As described previously, a non-surgical method that may have potential in relieving joint pain is through dietary supplements such as chondroitin and glucosamine. These products are specifically labeled as "dietary supplements" because the firms that manufacture or market these products do not claim medical efficacy. Promotional materials are typically vaguely worded with disclaimers that the purported effects have not been evaluated by the FDA. These products are not intended to diagnose, treat, cure, or prevent joint disease and are not drugs. A company that sells glucosamine, chondroitin, and other "nutraceuticals" (dietary supplements that may exhibit health benefits) may have anecdotal evidence or even peer-reviewed studies linking these products to certain conditions but may, nonetheless, lack the rigorous scientific studies required by the FDA to demonstrate significant effects.

As defined by the Federal Food, Drug, and Cosmetic Act [834], drugs are "articles intended for use in the diagnosis, cure, mitigation, treatment, or prevention of disease" and "articles (other than food) intended to affect the structure or any function of the body." The primary mode of drug effects is via chemical pathways. The Center for Drug Evaluation and Research (CDER) ensures that the nation's drugs are safe and effective. This includes over-the-counter, prescription, biological therapeutics, and generic drugs, as well as fluoride toothpaste, antiperspirants, and sunscreens. CDER's Division of Manufacturing and Product Quality ensures that drug manufacturers follow current good manufacturing practices (cGMP) as described by 21 CFR Parts 210 and 211 [835,836]. Adverse events are monitored by the Adverse Event Reporting System (AERS), a computerized information database designed to support the FDA's post-marketing safety surveillance program for all approved drug and therapeutic biologic products. While analgesics and COX-2 inhibitors, useful in relieving joint pain, are regulated by the CDER, the majority of tissue engineered products will likely not fall under CDER's jurisdiction.

Biological products, such as blood, vaccines, allergenics, tissues, and cellular and gene therapies that are derived from living sources (such as humans, animals, and microorganisms), are regulated by the Center for Biologics Evaluation and Research (CBER). Similar to drugs, manufacturers of biological products must follow cGMP as described by 21 CFR Part 211 [836] and report adverse events to AERS. Looking back on the tissue engineering methods described in this volume, many potential products can fall under this category. All autologous, allogeneic, and xenogeneic products are produced by cells and can potentially be regulated by CBER. It is important to note, though, that the primary mode of the effects of biologics is metabolic. Thus, ACI, a procedure that requires manipulation of the autologous cells that will metabolize and produce tissue *in situ*, is regulated by CBER.

A Biologics License Application (BLA) must be filed for the permission to introduce a biologic product into the market, and the BLA is regulated under 21 CFR 600 to 680. Applicants for a license must comply with requirements set forth by Form 356h, which includes the applicant information, product and manufacturing information, pre-clinical studies, clinical studies, and labeling. Of substantial burden are clinical studies, which can be slow and costly. For implants, the FDA has had a history of classifying most orthopaedic implants as medical devices, which have their own regulatory Center. In the case of ACI, the resulting implant is expected to generate tissue through metabolic means and, as a cell slurry, cannot withstand loading as an implant would and, thus, probably cannot be classified as a device. Applying for approval as a device, e.g., through the 510(k), may be less burdensome as explained in the next section.

Firms that manufacture, repackage, relabel, and/or import medical devices sold in the United States are subject to regulation by the Center for Devices and Radiological Health (CDRH). (As an aside, CDRH also regulates radiation-emitting electronic products, including x-ray systems, ultrasound equipment, microwave ovens and color televisions.) A medical device is "an instrument, apparatus, implement, machine, contrivance, implant, *in vitro* reagent, or other similar or related article, including any component, part, or accessory, which is intended for use in the diagnosis of disease or other conditions, or in the cure, mitigation, treatment, or prevention of disease" [834]. Tissue engineered cartilage, as an implant, may be regulated by CDRH even if it contains biologics or drugs, depending on the assignment by the Office of Combination Products. The primary mode of effect for devices is mechanical or electrical. Implants whose primary function is to bear mechanical load (as tissue engineered cartilage is designed to function) may have CDRH as its primary regulator. Devices regulated within the CDRH fall into three classes and have different requirements that the manufacturer must fulfill prior to introducing a product to market. As CDRH has been assigned jurisdiction over most orthopaedic implants, pathways to market through this Center will be described in the next section.

A medical product intended to promote public health must be both safe and effective. Scientific studies conducted to demonstrate both of these criteria can cost both time and money. The FDA, charged with reviewing the scientific data, must also devote similar resources. A shorter time-to-market can potentially result in the earlier realization in benefits to public health or in lack of adequate data to ensure safety. While excessive burdens in testing can deplete a company's resources and stifle product innovation, small clinical studies might not be able to detect rare adverse events associated with a device. Through its Centers and Offices, the FDA performs a balancing act between expediency and assurance of public welfare. To ease the regulatory burden of industry, FDA has established intercenter agreements, e.g., between the CBER and the CDRH, and between the CDER and the CDRH. Also, in order to ensure that adequate effort is expended in evaluating a device, the FDA Modernization Act of 1997 established three classes of devices, each requiring different degrees of rigor in demonstrating efficacy (all devices must be safe). Firms interested in obtaining approval for a device can also meet with the Office of Device Evaluation to determine the "least burdensome" method in showing that a device is effective, through the selection of appropriate pathways for the application.

5.5.2   DEVICE CLASSIFICATIONS AND PATHWAYS TO MARKET

Class I devices are low risk and pose minimal potential harm. The key regulatory compliance for a Class I device is that a company must demonstrate that it has implemented "general controls." General controls include quality system regulation (QSR), as described by 21 CFR 820 [837], to ensure adherence to predefined design controls and GMP, label requirements to prevent product mislabeling, and the use of Medical Device Reporting (MDR), not the AERS, as the mechanism to maintain records for the reporting of adverse events identified by the user, manufacturer, or distributer of the device. Hand-held instruments, elastic bandages, exam gloves, and enema kits are examples of Class I devices. As part of the pathway to market, medical devices must use forms FDA-2891 and FDA-2892 for establishment registration and medical device listing.

Aside from the requirements listed above for Class I devices, Class II devices, which are of moderate risk, often require a Premarket Notification 510(k) pathway to market. That is, before proposing to market a device, the manufacturer must notify the FDA 90 days in advance that the device is *substantially equivalent* to a predicate device legally in commercial distribution in the US before May 28, 1976. Substantially equivalent devices will generally be labeled as a Class I or II device. Applying for a 510(k) starts with the determination of an appropriate product code for the device and locating a predicate device from there. The FDA provides guidance documents that a company can review, and either a traditional, special, or abbreviated 510(k) may be filed. A traditional 510(k) takes about 90 days to review. A special 510(k) can be filed for when a device is modified. No changes to the intended use of the device can be made and the manufacturer must declare conformance to design controls. Review for a special 510(k) is faster, generally 30 days. An abbreviated 510(k) also has a reduced review time but not necessarily down to 30 days. It relies on use of guidance documents or special controls to provide a summary report that describes adherence to the relevant guidance document. After a Class II device goes to market, it may require special controls, such as postmarket surveillance, patient registries, guidances, and standards. Most joint arthroplasty components are approved as Class II devices, as well as other implanted materials such as pedicle screws and intramedullary nails.

Class III devices support or sustain human life, are of substantial importance in preventing impairment of human health, or present a potential, unreasonable risk of illness or injury. Devices for which substantially equivalent predicates are non-existent are also classified as Class III devices. If a new device is deemed to be substantially equivalent to a predicate Class III device, then it, too, will be a Class III device. Lastly, if a new device is determined to be substantially equivalent to a Class I or II device that was developed after 1976, it also automatically falls into Class III. A company may petition to have a new device reclassified to Class I or Class II. For a Class III device, a company must submit a premarket approval application (PMA) or Product Development Protocol (PDP) before legal distribution can occur. The PMA process will include both preclinical and clinical data to demonstrate safety and efficacy. In order to collect clinical data, a new device must first have an investigational device exemption (IDE) (see 21 CFR 812 [838]) before it can be used in humans. If the device has significant risk, the FDA and Institutional Review Board (IRB) must approve the study before initiation. If the device is of non-significant risk, only the IRB needs to approve.

Additional pathways to market include the Humanitarian Device Exemption (HDE) and the PDP (for Class III devices). HDE devices cannot have a profit margin as this pathway is intended for the development of devices to treat rare (< 4,000 patients per year) conditions. To speed the time to approval, regulatory burdens on the manufacturer are lessened for HDE devices as such devices need not have demonstrated efficacy. However, HDE devices do need to be approved by the IRB where they are used, and surgeons must be aware that efficacy has not yet been shown for these devices. A PDP is an alternative to the PMA. To purse this pathway, a company would work with the FDA in designing preclinical and clinical studies, protocols, assessment methods, and acceptance criteria. Few products have been approved through this pathway, though it may be speedier than a PMA as the FDA would be involved in the initial development of the product all the way to market.

Tissue engineered cartilage, though derived from biological sources, may be regulated as an implant instead and qualify for the shorter and less costly 510(k) pathway to market (instead of a BLA) if it is shown to be substantially equivalent to a predicate device (e.g., arthroplasty components). Due to the combinatorial nature of many tissue repair products, the Office of Combination Products may assign primary jurisdiction to the CDRH if a product's primary mode of effect is not metabolic. For example, Medtronic's InFuse Bone Graft/LT-Cage, consisting of a collagen scaffold with recombinant human BMP (rhBMP) enclosed within a metallic lumbar fusion device, is regulated as a medical device. In this case, the rhBMP is derived from a biological source, but the primary mode of action for the product is for mechanical support. Similarly, biologic products such as bone void fillers and demineralized bone matrix are classified as Class II devices by the FDA, while BMPs are classified as Class III devices. Depending on the nature of a tissue engineered product, whether it will be assigned to CBER or CDRH is something researchers and companies should consider as they develop the product because the pathways to market are substantially different for each Center. Furthermore, if a company would like to be regulated under CDRH, it should have a plan for which class a product falls into and project the time and financial burdens from there. At present, no tissue engineered implants are on the market, and it is likely for either CBER or CDRH to be involved. Companies that are traditionally used to pathways required by CDRH might start acquainting themselves with CBER regulations.

Getting a product approved by the FDA does not mean that it will automatically be approved for reimbursement by various private and public insurers. As engineered cartilage will be of use for the elderly, it is relevant to understand how Medicare approves of reimbursement. The Centers for Medicare & Medicaid Services (CMS) makes a decision based on whether a treatment is "reasonable and necessary." While criteria for "reasonable and necessary" are ill-defined, it is clear that the FDA's mandate of "safe and effective" is not sufficient in warranting reimbursement. Criteria for reimbursement by private insurers differ from company to company. While it is clear that a need for long-term cartilage resurfacing exists, whether tissue engineering will be a profitable technology in addressing this void will depend on many factors.

5.5.3   CURRENTLY AVAILABLE PRODUCTS

While articular cartilage tissue engineered products are currently not available in the US, several products available in Europe and others that are undergoing clinical trials in the US can serve as models to inform those seeking to develop and market tissue engineered articular cartilage. Regulatory processes in Europe are different from the US FDA, and are reviewed elsewhere [839, 840]. Engineered cartilage employing allogeneic cells may find commonalities with tissue explant allografts. For lesions greater than 2 cm, Regeneration Technologies has fresh-stored osteochondral allografts that are cleaned, processed, and preserved with maintained chondrocyte viability. A proprietary antibiotic soak and 14 days of culture monitoring (with 28 days of fungal culture monitoring) prepare the allografts for implantation [841].

Several examples of cartilage treatments using only a biomaterial are available as products. Trufit (Smith & Nephew) is a product based on biodegradable polymers and used world-wide. Other products in use include Gelrin by Regentis Biomaterials, a fibrin/polyethylene glycol hydrogel that can be cross-linked *in situ* for cartilage defects. BioPoly RS, a subsidiary of Schwartz Biomedical, developed the BioPoly RS (ReSurfacing) device, which is a hydrophilic polymer. SaluMedica has SaluCartilage, which is a hydrogel that can be used for cartilage damage with CE Mark approval in Europe but not FDA approval. BST-CarGel, by BioSyntech, is another product with CE Mark approval and is used for focal cartilage lesions. Results from clinical data on 40 subjects enrolled in a Canadian-European trial using BST-CarGel are currently being analyzed, and final clinical results from an 80-patient study using this product are anticipated in 2010. Kensey Nash is in a late pre-clinical study of an implant, composed of several biomaterials and a biologically active protein, for osteochondral defects.

Cell seeded scaffolds have also been approved for use in Europe. CaReS (Cartilage Regeneration System) by Arthro Kinetics, uses articular chondrocytes for transplantation. Arthro Kinetics also developed CartiPlug, which is an acellular collagen matrix for the same indication, and is currently developing CaReS Plus, which would contain cells for larger cartilage defects. BioTissue Technologies has BioSeed-C, an autologous 3D chondrocyte graft. Since December 2001 BioSeed-C started on controlled trials in selected clinics. A four-year follow up study has recently been released showing stability of the regenerated tissue [750]. CellGenix markets CartiGro, which is autologous chondrocytes combined with Chondro-Gide collagen membrane (from Geistlick Biomaterials), and is distributed in Europe by Stryker EMEA. Fidia developed Hyalograft C, which uses Hyaff, a hyaluronic acid derivative, and autologous chondrocytes. DePuy's Cartilage Autograft Implantation System (CAIS) will harvest, seed onto a biomaterial, and implant the product all in one surgical procedure. DePuy has recently completed a multi-center randomized pilot study evaluating the safety and performance of CAIS. Histogenics licensed Angiotech's ChondroGEL biomaterial to use in combination with its proprietary Tissue Engineering Support System in growing a product known as NeoCart, which is currently near completion of Phase II trials. According to the company's website, Histogenics also has a scaffold product, VeriCart, for cartilage regeneration that was scheduled to begin Phase 1 clinical trials in late summer of 2008. It is worth noting that the only cell-based product available in the US does not employ a scaffold. Genzyme Biosurgery's Carticel consists of the deliverance of a slurry of expanded autologous chondrocytes. Japan Tissue Engineering has an autologous cell transplantation method similar to Genzyme's. Isto Technologies is collaborating with Zimmer to tissue engineer Neocartilage, which is also not scaffold-based.

5.6   CHAPTER CONCEPTS

- Due to the low availability of differentiated, autologous chondrocytes, alternative cell sources are investigated as the future of articular cartilage tissue engineering. Isolation, differentiation, and purification protocols for several stem cell types, both adult (e.g., MSCs, iPSCs, dermis derived, adipose derived) and embryonic, are developed.

- Parallels and commonalities in the differentiation of various stem cell populations to chondrocyte-like cells are currently in the discovery and application phase.

- Engineered cartilage can be assessed histologically, immunohistologically, biochemically, via non-destructive imaging, and biomechanically in determining its functionality pre-implantation.

- One of the most important properties of engineered cartilage is compressive stiffness, but, with the existence of a myriad of protocols in determining this property, inter-study comparisons can sometimes be difficult. Consensus standards in this area, along with tensile, shear, friction, and fatigue testing will aid in the future development of an engineered cartilage product.

- Integration of the engineered cartilage can be a significant hurdle due to the inherent non-adhesive nature of the tissue.

- Both surgical and non-surgical methods exist for the management of cartilage injuries. Non-surgical methods can include supplements and drugs to reduce inflammation, although their efficacy may include artifacts that result in a placebo effect. Surgical methods include debridement, microfracture, and various transplants and implants.

- Allogeneic cartilage transplants have been applied with substantial clinical success, and immune reactions are often mild or absent.

- Xenogeneic transplants are currently examined in animals. The presence of an α-galactosyl carbohydrate structure in animals, but not in humans, results in an immune reaction. Elimination or reduction of the immune reaction is currently sought through the removal of the α-galactosyl carbohydrate structure and through decellularization of xenogeneic implants.

• The FDA regulates food, drug, and medical device safety through several Centers and Offices. As compliance requirements differ with each Center, it is important for a tissue engineering company to consider which pathway to market a product might take.

• Most orthopaedic implants have been classified as devices. Pathways to market for devices include filing for the 510(k), PMA, PDP, or HDE applications. While several cell and biologic products are available for cartilage resurfacing in Europe, few have been approved in the US.

Bibliography

[1] Guilak, F., L.A. Setton, and V.B. Kraus, *Structure and function of articular cartilage*, in *Principles And Practice Of Orthopaedic Sports Medicine.*, W.E. Garrett, et al., Editors. 2000, Lippincott Williams & Wilkins: Philadelphia, PA.

[2] Wu, W., et al., *Sites of collagenase cleavage and denaturation of type II collagen in aging and osteoarthritic articular cartilage and their relationship to the distribution of matrix metalloproteinase 1 and matrix metalloproteinase 13.* Arthritis Rheum, 2002. **46**(8): p. 2087-94. DOI: 10.1002/art.10428

[3] Mankin, H.J. and L. Lippiello, *The turnover of adult rabbit articular cartilage.* J Bone Joint Surg Am, 1969. **51**(8): p. 1591-600.

[4] Darling, E.M. and K.A. Athanasiou, *Articular cartilage bioreactors and bioprocesses [published erratum appears in Tissue Eng 2003 9(3): 565].* Tissue Eng, 2003. **9**(1): p. 9-26. DOI: 10.1089/107632703762687492

[5] O'Hara, B.P., J.P. Urban, and A. Maroudas, *Influence of cyclic loading on the nutrition of articular cartilage.* Ann Rheum Dis, 1990. **49**(7): p. 536-9. DOI: 10.1136/ard.49.7.536

[6] Linn, F.C. and L. Sokoloff, *Movement and Composition of Interstitial Fluid of Cartilage.* Arthritis Rheum, 1965. **8**: p. 481-94. DOI: 10.1002/art.1780080402

[7] Ateshian, G.A., et al., *An asymptotic solution for the contact of two biphasic cartilage layers.* J Biomech, 1994. **27**(11): p. 1347-60. DOI: 10.1016/0021-9290(94)90044-2

[8] Mow, V.C., et al., *Biphasic indentation of articular cartilage–II. A numerical algorithm and an experimental study.* J Biomech, 1989. **22**(8-9): p. 853-61. DOI: 10.1016/0021-9290(89)90069-9

[9] Deshmukh, K. and M.E. Nimni, *Isolation and characterization of cyanogen bromide peptides from the collagen of bovine articular cartilage.* Biochem J, 1973. **133**(4): p. 615-22.

[10] Clark, J.M., *The organisation of collagen fibrils in the superficial zones of articular cartilage.* J Anat, 1990. **171**: p. 117-30.

[11] Eyre, D.R. and J.J. Wu, *Collagen structure and cartilage matrix integrity.* J Rheumatol Suppl, 1995. **43**: p. 82-5.

[12] Responte, D.J., R.M. Natoli, and K.A. Athanasiou, *Collagens of articular cartilage: structure, function, and importance in tissue engineering.* Crit Rev Biomed Eng, 2007. **35**(5): p. 363-411.

[13] Choi, J.B., et al., *Zonal changes in the three-dimensional morphology of the chondron under compression: the relationship among cellular, pericellular, and extracellular deformation in articular cartilage.* J Biomech, 2007. **40**(12): p. 2596-603. DOI: 10.1016/j.jbiomech.2007.01.009

[14] McDevitt, C.A., J. Marcelino, and L. Tucker, *Interaction of intact type VI collagen with hyaluronan.* FEBS Lett, 1991. **294**(3): p. 167-70. DOI: 10.1016/0014-5793(91)80660-U

[15] Aigner, T., et al., *Type X collagen expression in osteoarthritic and rheumatoid articular cartilage.* Virchows Arch B Cell Pathol Incl Mol Pathol, 1993. **63**(4): p. 205-11.

[16] Hardingham, T.E., A.J. Fosang, and J. Dudhia, *The structure, function and turnover of aggrecan, the large aggregating proteoglycan from cartilage.* Eur J Clin Chem Clin Biochem, 1994. **32**(4): p. 249-57.

[17] Roughley, P.J. and E.R. Lee, *Cartilage proteoglycans: structure and potential functions.* Microsc Res Tech, 1994. **28**(5): p. 385-97. DOI: 10.1002/jemt.1070280505

[18] Watanabe, H., Y. Yamada, and K. Kimata, *Roles of aggrecan, a large chondroitin sulfate proteoglycan, in cartilage structure and function.* J Biochem, 1998. **124**(4): p. 687-93.

[19] Maroudas, A., H. Muir, and J. Wingham, *The correlation of fixed negative charge with glycosaminoglycan content of human articular cartilage.* Biochim Biophys Acta, 1969. **177**(3): p. 492-500.

[20] Maroudas, A.I., *Balance between swelling pressure and collagen tension in normal and degenerate cartilage.* Nature, 1976. **260**(5554): p. 808-9.

[21] Flannery, C.R., et al., *Articular cartilage superficial zone protein (SZP) is homologous to megakaryocyte stimulating factor precursor and is a multifunctional proteoglycan with potential growth-promoting, cytoprotective, and lubricating properties in cartilage metabolism.* Biochem Biophys Res Commun, 1999. **254**(3): p. 535-41. DOI: 10.1006/bbrc.1998.0104

[22] Boskey, A.L., *Current concepts of the physiology and biochemistry of calcification.* Clinical Orthopaedics & Related Research, 1981(157): p. 225-57.

[23] Poole, A.R., et al., *Cartilage macromolecules and the calcification of cartilage matrix.* Anat Rec, 1989. **224**(2): p. 167-79.

[24] Pritzker, K.P., P.T. Cheng, and R.C. Renlund, *Calcium pyrophosphate crystal deposition in hyaline cartilage. Ultrastructural analysis and implications for pathogenesis.* J Rheumatol, 1988. **15**(5): p. 828-35.

[25] Heinegard, D. and A. Oldberg, *Structure and biology of cartilage and bone matrix noncollagenous macromolecules.* Faseb J, 1989. **3**(9): p. 2042-51.

[26] Hunziker, E.B., M. Michel, and D. Studer, *Ultrastructure of adult human articular cartilage matrix after cryotechnical processing.* Microsc Res Tech, 1997. **37**(4): p. 271-84. DOI: 10.1002/(SICI)1097-0029(19970515)37:4<271::AID-JEMT3>3.0.CO;2-O

[27] Poole, C.A., M.H. Flint, and B.W. Beaumont, *Chondrons in cartilage: ultrastructural analysis of the pericellular microenvironment in adult human articular cartilages.* J Orthop Res, 1987. **5**(4): p. 509-22. DOI: 10.1002/jor.1100050406

[28] Mac, C.M., *The movements of bones and joints; the mechanical structure of articulating cartilage.* J Bone Joint Surg Br, 1951. **33B**(2): p. 251-7.

[29] Kumar, P., et al., *Role of uppermost superficial surface layer of articular cartilage in the lubrication mechanism of joints.* J Anat, 2001. **199**(Pt 3): p. 241-50. DOI: 10.1046/j.1469-7580.2001.19930241.x

[30] Teeple, E., et al., *Frictional properties of Hartley guinea pig knees with and without proteolytic disruption of the articular surfaces.* Osteoarthritis Cartilage, 2007. **15**(3): p. 309-15. DOI: 10.1016/j.joca.2006.08.011

[31] Wu, J.P., T.B. Kirk, and M.H. Zheng, *Study of the collagen structure in the superficial zone and physiological state of articular cartilage using a 3D confocal imaging technique.* J Orthop Surg, 2008. **3**: p. 29. DOI: 10.1186/1749-799X-3-29

[32] Meachim, G. and S.R. Sheffield, *Surface ultrastructure of mature adult human articular cartilage.* J Bone Joint Surg Br, 1969. **51**(3): p. 529-39.

[33] Muir, H., P. Bullough, and A. Maroudas, *The distribution of collagen in human articular cartilage with some of its physiological implications.* J Bone Joint Surg Br, 1970. **52**(3): p. 554-63.

[34] Maroudas, A., *Physicochemical properties of articular cartilage*, in *Adult Articular Cartilage*, M. Freeman, Editor. 1979, Pitman Medical: Tunbridge Wells.

[35] Stockwell, R.A., *The interrelationship of cell density and cartilage thickness in mammalian articular cartilage.* J Anat, 1971. **109**(3): p. 411-21.

[36] Eggli, P.S., E.B. Hunziker, and R.K. Schenk, *Quantitation of structural features characterizing weight- and less- weight-bearing regions in articular cartilage: a stereological analysis of medial femoral condyles in young adult rabbits.* Anat Rec, 1988. **222**(3): p. 217-27. DOI: 10.1002/ar.1092220302

[37] Venn, M. and A. Maroudas, *Chemical composition and swelling of normal and osteoarthrotic femoral head cartilage. I. Chemical composition.* Ann Rheum Dis, 1977. **36**(2): p. 121-9. DOI: 10.1136/ard.36.2.121

[38] Redler, I., et al., *The ultrastructure and biomechanical significance of the tidemark of articular cartilage.* Clin Orthop Relat Res, 1975(112): p. 357-62.

[39] Radin, E.L. and R.M. Rose, *Role of subchondral bone in the initiation and progression of cartilage damage.* Clin Orthop Relat Res, 1986(213): p. 34-40.

[40] Hunziker, E.B., W. Herrmann, and R.K. Schenk, *Ruthenium hexammine trichloride (RHT)-mediated interaction between plasmalemmal components and pericellular matrix proteoglycans is responsible for the preservation of chondrocytic plasma membranes in situ during cartilage fixation.* J Histochem Cytochem, 1983. **31**(6): p. 717-27.

[41] Poole, C.A., S. Ayad, and J.R. Schofield, *Chondrons from articular cartilage: I. Immunolocalization of type VI collagen in the pericellular capsule of isolated canine tibial chondrons.* J Cell Sci, 1988. **90**(Pt 4): p. 635-43.

[42] Youn, I., et al., *Zonal variations in the three-dimensional morphology of the chondron measured in situ using confocal microscopy.* Osteoarthritis Cartilage, 2006. **14**(9): p. 889-97. DOI: 10.1016/j.joca.2006.02.017

[43] Mow, V.C., E.L. Flatow, and G.A. Ateshian, *Biomechanics*, in *Orthopaedic basic science: Biology and biomechanics of the musculoskeletal system*, J.A. Buckwalter, T.A. Einhorn, and S.R. Simon, Editors. 2000, American Academy of Orthopaedic Surgeons. p. 140-42.

[44] Hodge, W.A., et al., *Contact pressures from an instrumented hip endoprosthesis.* J Bone Joint Surg Am, 1989. **71**(9): p. 1378-86.

[45] Hayes, W.C. and L.F. Mockros, *Viscoelastic properties of human articular cartilage.* J Appl Physiol, 1971. **31**(4): p. 562-8.

[46] Mak, A.F., *The apparent viscoelastic behavior of articular cartilage–the contributions from the intrinsic matrix viscoelasticity and interstitial fluid flows.* J Biomech Eng, 1986. **108**(2): p. 123-30. DOI: 10.1115/1.3138591

[47] Setton, L.A., W. Zhu, and V.C. Mow, *The biphasic poroviscoelastic behavior of articular cartilage: role of the surface zone in governing the compressive behavior.* J Biomech, 1993. **26**(4-5): p. 581-92. DOI: 10.1016/0021-9290(93)90019-B

[48] Mow, V.C., M.H. Holmes, and W.M. Lai, *Fluid transport and mechanical properties of articular cartilage: a review.* J Biomech, 1984. **17**(5): p. 377-94. DOI: 10.1016/0021-9290(84)90031-9

[49] Schinagl, R.M., et al., *Depth-dependent confined compression modulus of full-thickness bovine articular cartilage.* J Orthop Res, 1997. **15**(4): p. 499-506. DOI: 10.1002/jor.1100150404

[50] Athanasiou, K.A., et al., *Interspecies comparisons of in situ intrinsic mechanical properties of distal femoral cartilage.* J Orthop Res, 1991. **9**(3): p. 330-40. DOI: 10.1002/jor.1100090304

[51] Athanasiou, K.A., A. Agarwal, and F.J. Dzida, *Comparative study of the intrinsic mechanical properties of the human acetabular and femoral head cartilage.* Journal of orthopaedic research : official publication of the Orthopaedic Research Society., 1994. **12**(3): p. 340-9. DOI: 10.1002/jor.1100120306

[52] Athanasiou, K.A., et al., *Biomechanical topography of human articular cartilage in the first metatarsophalangeal joint.* Clinical orthopaedics and related research., 1998. **348**: p. 269-81. DOI: 10.1097/00003086-199803000-00038

[53] Athanasiou, K.A., G.G. Niederauer, and R.C. Schenck, Jr., *Biomechanical topography of human ankle cartilage.* Annals of biomedical engineering., 1995. **23**(5): p. 697-704. DOI: 10.1007/BF02584467

[54] Maroudas, A. and P. Bullough, *Permeability of articular cartilage.* Nature, 1968. **219**(5160): p. 1260-1.

[55] Mansour, J.M. and V.C. Mow, *The permeability of articular cartilage under compressive strain and at high pressures.* J Bone Joint Surg Am, 1976. **58**(4): p. 509-16.

[56] Park, S., et al., *Cartilage interstitial fluid load support in unconfined compression.* J Biomech, 2003. **36**(12): p. 1785-96. DOI: 10.1016/S0021-9290(03)00231-8

[57] Lai, W.M., V.C. Mow, and V. Roth, *Effects of nonlinear strain-dependent permeability and rate of compression on the stress behavior of articular cartilage.* J Biomech Eng, 1981. **103**(2): p. 61-6. DOI: 10.1115/1.3138261

[58] Mow, V.C., et al., *Biphasic creep and stress relaxation of articular cartilage in compression? Theory and experiments.* J Biomech Eng, 1980. **102**(1): p. 73-84. DOI: 10.1115/1.3138202

[59] Armstrong, C.G., A.S. Bahrani, and D.L. Gardner, *In vitro measurement of articular cartilage deformations in the intact human hip joint under load.* J Bone Joint Surg Am, 1979. **61**(5): p. 744-55.

[60] Kempson, G.E., M.A. Freeman, and S.A. Swanson, *Tensile properties of articular cartilage.* Nature, 1968. **220**(5172): p. 1127-8.

[61] Roth, V. and V.C. Mow, *The intrinsic tensile behavior of the matrix of bovine articular cartilage and its variation with age.* J Bone Joint Surg Am, 1980. **62**(7): p. 1102-17.

[62] Woo, S.L., et al., *Large deformation nonhomogeneous and directional properties of articular cartilage in uniaxial tension.* J Biomech, 1979. **12**(6): p. 437-46. DOI: 10.1016/0021-9290(79)90028-9

[63] Akizuki, S., et al., *Tensile properties of human knee joint cartilage: I. Influence of ionic conditions, weight bearing, and fibrillation on the tensile modulus.* J Orthop Res, 1986. **4**(4): p. 379-92. DOI: 10.1002/jor.1100040401

[64] Kempson, G.E., et al., *The tensile properties of the cartilage of human femoral condyles related to the content of collagen and glycosaminoglycans.* Biochim Biophys Acta, 1973. **297**(2): p. 456-72.

[65] Schmidt, M.B., et al., *Effects of proteoglycan extraction on the tensile behavior of articular cartilage.* J Orthop Res, 1990. **8**(3): p. 353-63. DOI: 10.1002/jor.1100080307

[66] Woo, S.L., W.H. Akeson, and G.F. Jemmott, *Measurements of nonhomogeneous, directional mechanical properties of articular cartilage in tension.* J Biomech, 1976. **9**(12): p. 785-91. DOI: 10.1016/0021-9290(76)90186-X

[67] Abbot, A.E., W.N. Levine, and V.C. Mow, *Biomechanics of the Articular Cartilage and Menisci of the Adult Knee* in *The Adult Knee*, J.J. Callaghan, et al., Editors. 2003, Lippincott Williams & Wilkins: Philadelphia, PA.

[68] Setton, L.A., V.C. Mow, and D.S. Howell, *Mechanical behavior of articular cartilage in shear is altered by transection of the anterior cruciate ligament.* J Orthop Res, 1995. **13**(4): p. 473-82. DOI: 10.1002/jor.1100130402

[69] Zhu, W., et al., *Viscoelastic shear properties of articular cartilage and the effects of glycosidase treatments.* J Orthop Res, 1993. **11**(6): p. 771-81. DOI: 10.1002/jor.1100110602

[70] Simon, W.H., A. Mak, and A. Spirt, *The effect of shear fatigue on bovine articular cartilage.* J Orthop Res, 1990. **8**(1): p. 86-93. DOI: 10.1002/jor.1100080111

[71] Hlavacek, M., *The role of synovial fluid filtration by cartilage in lubrication of synovial joints–II. Squeeze-film lubrication: homogeneous filtration.* J Biomech, 1993. **26**(10): p. 1151-60. DOI: 10.1016/0021-9290(93)90063-K

[72] Hou, J.S., et al., *An analysis of the squeeze-film lubrication mechanism for articular cartilage.* J Biomech, 1992. **25**(3): p. 247-59. DOI: 10.1016/0021-9290(92)90024-U

[73] Dowson, D. and Z.M. Jin, *Micro-elastohydrodynamic lubrication of synovial joints.* Eng Med, 1986. **15**(2): p. 63-5.

[74] Suciu, A.N., T. Iwatsubo, and M. Matsuda, *Theoretical investigation of an artificial joint with micro-pocket-covered component and biphasic cartilage on the opposite articulating surface.* J Biomech Eng, 2003. **125**(4): p. 425-33. DOI: 10.1115/1.1589505

[75] Charnley, J., *The lubrication of animal joints in relation to surgical reconstruction by arthroplasty.* Ann Rheum Dis, 1960. **19**: p. 10-9. DOI: 10.1136/ard.19.1.10

[76] Wright, V. and D. Dowson, *Lubrication and cartilage.* J Anat, 1976. **121**(Pt 1): p. 107-18.

[77] Coles, J.M., et al., *In situ friction measurement on murine cartilage by atomic force microscopy.* J Biomech, 2008. **41**(3): p. 541-8. DOI: 10.1016/j.jbiomech.2007.10.013

[78] Krishnan, R., M. Kopacz, and G.A. Ateshian, *Experimental verification of the role of interstitial fluid pressurization in cartilage lubrication.* J Orthop Res, 2004. **22**(3): p. 565-70. DOI: 10.1016/j.orthres.2003.07.002

[79] McCutchen, C.W., *The frictional properties of animal joints.* Wear, 1962. **5**(1): p. 1-17. DOI: 10.1016/0043-1648(62)90176-X

[80] Ateshian, G.A., H. Wang, and W.M. Lai, *The role of interstitial fluid pressurization and surface porosities on the boundary friction of articular cartilage.* J Tribol, 1998. **120**: p. 241-8. DOI: 10.1115/1.2834418

[81] Forster, H. and J. Fisher, *The influence of loading time and lubricant on the friction of articular cartilage.* Proc Inst Mech Eng [H], 1996. **210**(2): p. 109-19.

[82] Mow, V.C. and G.A. Ateshian, *Lubrication and wear of diarthrodial joints,* in *Basic Orthopaedic Biomechanics,* V.C. Mow and W.C. Hayes, Editors. 1997, Lippincott-Raven: Philadelphia.

[83] Aydelotte, M.B. and K.E. Kuettner, *Differences between sub-populations of cultured bovine articular chondrocytes. I. Morphology and cartilage matrix production.* Connect Tissue Res, 1988. **18**(3): p. 205-22. DOI: 10.3109/03008208809016808

[84] Darling, E.M., J.C.Y. Hu, and K.A. Athanasiou, *Zonal and topographical differences in articular chondrocyte gene expression.* J Orthop Res, 2004. **22**(6): p. 1182-1187. DOI: 10.1016/j.orthres.2004.03.001

[85] Darling, E.M., S. Zauscher, and F. Guilak, *Viscoelastic properties of zonal articular chondrocytes measured by atomic force microscopy.* Osteoarthritis Cartilage, 2006. **14**(6): p. 571-9. DOI: 10.1016/j.joca.2005.12.003

[86] Darling, E.M. and K.A. Athanasiou, *Growth factor impact on articular cartilage subpopulations.* Cell Tissue Res, 2005: p. 1-11. DOI: 10.1007/s00441-005-0020-4

[87] Archer, C.W., et al., *Phenotypic modulation in sub-populations of human articular chondrocytes in vitro.* J Cell Sci, 1990. **97**(Pt 2): p. 361-71.

[88] Trippel, S.B., et al., *Characterization of chondrocytes from bovine articular cartilage: I. Metabolic and morphological experimental studies.* J Bone Joint Surg [Am], 1980. **62**(5): p. 816-20.

[89] Aydelotte, M.B., R.R. Greenhill, and K.E. Kuettner, *Differences between sub-populations of cultured bovine articular chondrocytes. II. Proteoglycan metabolism.* Connect Tissue Res, 1988. **18**(3): p. 223-34. DOI: 10.3109/03008208809016808

[90] Siczkowski, M. and F.M. Watt, *Subpopulations of chondrocytes from different zones of pig articular cartilage. Isolation, growth and proteoglycan synthesis in culture.* J Cell Sci, 1990. **97**(Pt 2): p. 349-60.

[91] Zanetti, M., A. Ratcliffe, and F.M. Watt, *Two subpopulations of differentiated chondrocytes identified with a monoclonal antibody to keratan sulfate.* J Cell Biol, 1985. **101**(1): p. 53-9. DOI: 10.1083/jcb.101.1.53

[92] Durrant, L.A., et al., *Organisation of the chondrocyte cytoskeleton and its response to changing mechanical conditions in organ culture.* J Anat, 1999. **194**(Pt 3): p. 343-53. DOI: 10.1046/j.1469-7580.1999.19430343.x

[93] Ralphs, J.R., R.N. Tyers, and M. Benjamin, *Development of functionally distinct fibrocartilages at two sites in the quadriceps tendon of the rat: the suprapatella and the attachment to the patella.* Anat Embryol, 1992. **185**(2): p. 181-7. DOI: 10.1007/BF00185920

[94] Ghadially, F.N., *Fine Structure of Synovial Joints: A Text and Atlas of the Ultrastructure of Normal and Pathological Articular Tissues.* 1983, London: Butterworths.

[95] Poole, A.R., et al., *Osteoarthritis in the human knee: a dynamic process of cartilage matrix degradation, synthesis and reorganization.* Agents Actions Suppl, 1993. **39**: p. 3-13.

[96] Darling, E.M. and K.A. Athanasiou, *Rapid phenotypic changes in passaged articular chondrocyte subpopulations.* J Orthop Res, 2005. **23**(2): p. 425-32. DOI: 10.1016/j.orthres.2004.08.008

[97] Darling, E.M. and K.A. Athanasiou, *Retaining zonal chondrocyte phenotype by means of novel growth environments.* Tissue Eng, 2005. **11**(3-4): p. 395-403. DOI: 10.1089/ten.2005.11.395

[98] Mow, V.C., et al., *Stress, strain, pressure, and flow fields in articular cartilage,* in *Cell Mechanics and Cellular Engineering,* V.C. Mow, et al., Editors. 1994, Springer Verlag: New York. p. 345-379.

[99] Athanasiou, K., et al., *Tissue Engineering of Temporomandibular Joint Cartilage.* Synthesis Lectures on Tissue Engineering, ed. K. Athanasiou. Vol. 1. 2009: Morgan & Claypool Publishers. 122.

[100] Cserjesi, P., et al., *Scleraxis: a basic helix-loop-helix protein that prefigures skeletal formation during mouse embryogenesis.* Development, 1995. **121**(4): p. 1099-110.

[101] Sosic, D., et al., *Regulation of paraxis expression and somite formation by ectoderm- and neural tube-derived signals.* Dev Biol, 1997. **185**(2): p. 229-43. DOI: 10.1006/dbio.1997.8561

[102] Roberts, D.J., et al., *Sonic hedgehog is an endodermal signal inducing Bmp-4 and Hox genes during induction and regionalization of the chick hindgut.* Development, 1995. **121**(10): p. 3163-74.

[103] Lengerke, C., et al., *BMP and Wnt specify hematopoietic fate by activation of the Cdx-Hox pathway.* Cell Stem Cell, 2008. **2**(1): p. 72-82. DOI: 10.1016/j.stem.2007.10.022

[104] Li, X. and X. Cao, *BMP signaling and HOX transcription factors in limb development.* Front Biosci, 2003. **8**: p. s805-12. DOI: 10.2741/1150

[105] Tucker, A.S., A. Al Khamis, and P.T. Sharpe, *Interactions between Bmp-4 and Msx-1 act to restrict gene expression to odontogenic mesenchyme.* Dev Dyn, 1998. **212**(4): p. 533-9. DOI: 10.1002/(SICI)1097-0177(199808)212:4<533::AID-AJA6>3.0.CO;2-I

[106] Tureckova, J., et al., *Comparison of expression of the msx-1, msx-2, BMP-2 and BMP-4 genes in the mouse upper diastemal and molar tooth primordia.* Int J Dev Biol, 1995. **39**(3): p. 459-68.

[107] Chiquet-Ehrismann, R., P. Kalla, and C.A. Pearson, *Participation of tenascin and transforming growth factor-beta in reciprocal epithelial-mesenchymal interactions of MCF7 cells and fibroblasts.* Cancer Res, 1989. **49**(15): p. 4322-5.

[108] Vaahtokari, A., S. Vainio, and I. Thesleff, *Associations between transforming growth factor beta 1 RNA expression and epithelial-mesenchymal interactions during tooth morphogenesis.* Development, 1991. **113**(3): p. 985-94.

[109] Leonard, C.M., et al., *Role of transforming growth factor-beta in chondrogenic pattern formation in the embryonic limb: stimulation of mesenchymal condensation and fibronectin gene expression by exogenenous TGF-beta and evidence for endogenous TGF-beta-like activity.* Dev Biol, 1991. **145**(1): p. 99-109. DOI: 10.1016/0012-1606(91)90216-P

[110] Chimal-Monroy, J. and L. Diaz de Leon, *Expression of N-cadherin, N-CAM, fibronectin and tenascin is stimulated by TGF-beta1, beta2, beta3 and beta5 during the formation of precartilage condensations.* Int J Dev Biol, 1999. **43**(1): p. 59-67.

[111] Oberlender, S.A. and R.S. Tuan, *Expression and functional involvement of N-cadherin in embryonic limb chondrogenesis.* Development, 1994. **120**(1): p. 177-87.

[112] Tavella, S., et al., *N-CAM and N-cadherin expression during in vitro chondrogenesis.* Exp Cell Res, 1994. **215**(2): p. 354-62. DOI: 10.1006/excr.1994.1352

[113] Woodward, W.A. and R.S. Tuan, *N-Cadherin expression and signaling in limb mesenchymal chondrogenesis: stimulation by poly-L-lysine.* Dev Genet, 1999. **24**(1-2): p. 178-87. DOI: 10.1002/(SICI)1520-6408(1999)24:1/2<178::AID-DVG16>3.0.CO;2-M

[114] Luo, Y., I. Kostetskii, and G.L. Radice, *N-cadherin is not essential for limb mesenchymal chondrogenesis.* Dev Dyn, 2005. **232**(2): p. 336-44. DOI: 10.1002/dvdy.20241

[115] Salmivirta, M., et al., *Syndecan from embryonic tooth mesenchyme binds tenascin.* J Biol Chem, 1991. **266**(12): p. 7733-9.

[116] Bernfield, M. and R.D. Sanderson, *Syndecan, a developmentally regulated cell surface proteoglycan that binds extracellular matrix and growth factors.* Philos Trans R Soc Lond B Biol Sci, 1990. **327**(1239): p. 171-86.

[117] Capdevila, J. and R.L. Johnson, *Endogenous and ectopic expression of noggin suggests a conserved mechanism for regulation of BMP function during limb and somite patterning.* Dev Biol, 1998. **197**(2): p. 205-17. DOI: 10.1006/dbio.1997.8824

[118] McMahon, J.A., et al., *Noggin-mediated antagonism of BMP signaling is required for growth and patterning of the neural tube and somite.* Genes Dev, 1998. **12**(10): p. 1438-52. DOI: 10.1101/gad.12.10.1438

[119] Goff, D.J. and C.J. Tabin, *Analysis of Hoxd-13 and Hoxd-11 misexpression in chick limb buds reveals that Hox genes affect both bone condensation and growth.* Development, 1997. **124**(3): p. 627-36.

[120] Hall, B.K., *Bones and Cartilage: Developmental Skeletal Biology.* 1st ed. 2005: Academic Press.

[121] Roark, E.F. and K. Greer, *Transforming growth factor-beta and bone morphogenetic protein-2 act by distinct mechanisms to promote chick limb cartilage differentiation in vitro.* Dev Dyn, 1994. **200**(2): p. 103-16.

[122] Horton, W.A., *The biology of bone growth.* Growth Genet. Horm., 1990. **6**(2): p. 1-3.

[123] Farnum, C.E., et al., *Volume increase in growth plate chondrocytes during hypertrophy: the contribution of organic osmolytes.* Bone, 2002. **30**(4): p. 574-81. DOI: 10.1016/S8756-3282(01)00710-4

[124] Descalzi Cancedda, F., et al., *Production of angiogenesis inhibitors and stimulators is modulated by cultured growth plate chondrocytes during in vitro differentiation: dependence on extracellular matrix assembly.* Eur J Cell Biol, 1995. **66**(1): p. 60-8.

[125] Gilbert, S.F., *Developmental Biology.* Seventh ed. 2003, Sunderland, MA: Sinauer Associates, Inc., Publishers.

[126] Thompson, T.J., P.D. Owens, and D.J. Wilson, *Intramembranous osteogenesis and angiogenesis in the chick embryo.* J Anat, 1989. **166**: p. 55-65.

[127] Shapiro, I.M., et al., *Developmental regulation of creatine kinase activity in cells of the epiphyseal growth cartilage.* J Bone Miner Res, 1992. **7**(5): p. 493-500.

[128] Gitelman, S.E., et al., *Recombinant Vgr-1/BMP-6-expressing tumors induce fibrosis and endochondral bone formation in vivo.* J Cell Biol, 1994. **126**(6): p. 1595-609. DOI: 10.1083/jcb.126.6.1595

[129] Houston, B., B.H. Thorp, and D.W. Burt, *Molecular cloning and expression of bone morphogenetic protein-7 in the chick epiphyseal growth plate.* J Mol Endocrinol, 1994. **13**(3): p. 289-301. DOI: 10.1677/jme.0.0130289

[130] De Luca, F., et al., *Regulation of growth plate chondrogenesis by bone morphogenetic protein-2.* Endocrinology, 2001. **142**(1): p. 430-6.

[131] Shum, L., et al., *BMP4 promotes chondrocyte proliferation and hypertrophy in the endochondral cranial base.* Int J Dev Biol, 2003. **47**(6): p. 423-31.

[132] Wang, G. and F. Beier, *Rac1/Cdc42 and RhoA GTPases antagonistically regulate chondrocyte proliferation, hypertrophy, and apoptosis.* J Bone Miner Res, 2005. **20**(6): p. 1022-31. DOI: 10.1359/JBMR.050113

[133] Wu, Q.Q. and Q. Chen, *Mechanoregulation of chondrocyte proliferation, maturation, and hypertrophy: ion-channel dependent transduction of matrix deformation signals.* Exp Cell Res, 2000. **256**(2): p. 383-91. DOI: 10.1006/excr.2000.4847

[134] Reginato, A.M., et al., *Effects of calcium deficiency on chondrocyte hypertrophy and type X collagen expression in chick embryonic sternum.* Dev Dyn, 1993. **198**(4): p. 284-95.

[135] Yang, X., et al., *TGF-beta/Smad3 signals repress chondrocyte hypertrophic differentiation and are required for maintaining articular cartilage.* J Cell Biol, 2001. **153**(1): p. 35-46. DOI: 10.1083/jcb.153.1.35

[136] Mancilla, E.E., et al., *Effects of fibroblast growth factor-2 on longitudinal bone growth.* Endocrinology, 1998. **139**(6): p. 2900-4. DOI: 10.1210/en.139.6.2900

[137] Johnson, K.A., D.M. Rose, and R.A. Terkeltaub, *Factor XIIIA mobilizes transglutaminase 2 to induce chondrocyte hypertrophic differentiation.* J Cell Sci, 2008. **121**(Pt 13): p. 2256-64. DOI: 10.1242/jcs.011262

[138] Haaijman, A., et al., *OP-1 (BMP-7) affects mRNA expression of type I, II, X collagen, and matrix Gla protein in ossifying long bones in vitro.* J Bone Miner Res, 1997. **12**(11): p. 1815-23. DOI: 10.1359/jbmr.1997.12.11.1815

[139] O'Keefe, R.J., et al., *Differential regulation of type-II and type-X collagen synthesis by parathyroid hormone-related protein in chick growth-plate chondrocytes.* J Orthop Res, 1997. **15**(2): p. 162-74. DOI: 10.1002/jor.1100150203

[140] Horiki, M., et al., *Smad6/Smurf1 overexpression in cartilage delays chondrocyte hypertrophy and causes dwarfism with osteopenia.* J Cell Biol, 2004. **165**(3): p. 433-45. DOI: 10.1083/jcb.200311015

[141] Kavumpurath, S. and B.K. Hall, *Lack of either chondrocyte hypertrophy or osteogenesis in Meckel's cartilage of the embryonic chick exposed to epithelia and to thyroxine in vitro.* J Craniofac Genet Dev Biol, 1990. **10**(3): p. 263-75.

[142] Kandel, R.A., et al., *In vitro formation of mineralized cartilagenous tissue by articular chondrocytes.* In Vitro Cell Dev Biol Anim, 1997. **33**(3): p. 174-81. DOI: 10.1007/s11626-997-0138-7

[143] Armstrong, C.G. and D.L. Gardner, *Thickness and distribution of human femoral head articular cartilage. Changes with age.* Ann Rheum Dis, 1977. **36**(5): p. 407-12. DOI: 10.1136/ard.36.5.407

[144] Meachim, G., *Effect of age on the thickness of adult articular cartilage at he shoulder joint.* Ann Rheum Dis, 1971. **30**(1): p. 43-6. DOI: 10.1136/ard.30.1.43

[145] Meachim, G., G. Bentley, and R. Baker, *Effect of age on thickness of adult patellar articular cartilage.* Ann Rheum Dis, 1977. **36**(6): p. 563-8. DOI: 10.1136/ard.36.6.563

[146] Vignon, E., M. Arlot, and G. Vignon, *[The cellularity of fibrillated articular cartilage. A comparative study of age-related and osteoarthrotic cartilage lesions from the human femoral head].* Pathol Biol (Paris), 1977. **25**(1): p. 29-32.

[147] Adams, C.S. and W.E. Horton, Jr., *Chondrocyte apoptosis increases with age in the articular cartilage of adult animals.* Anat Rec, 1998. **250**(4): p. 418-25. DOI: 10.1002/(SICI)1097-0185(199804)250:4<418::AID-AR4>3.0.CO;2-T

[148] Barbero, A., et al., *Age related changes in human articular chondrocyte yield, proliferation and post-expansion chondrogenic capacity.* Osteoarthritis Cartilage, 2004. **12**(6): p. 476-84. DOI: 10.1016/j.joca.2004.02.010

[149] Martin, J.A., S.M. Ellerbroek, and J.A. Buckwalter, *Age-related decline in chondrocyte response to insulin-like growth factor-I: the role of growth factor binding proteins.* J Orthop Res, 1997. **15**(4): p. 491-8. DOI: 10.1002/jor.1100150403

[150] Eyre, D.R., I.R. Dickson, and K. Van Ness, *Collagen cross-linking in human bone and articular cartilage. Age-related changes in the content of mature hydroxypyridinium residues.* Biochem J, 1988. **252**(2): p. 495-500.

[151] Bank, R.A., et al., *Ageing and zonal variation in post-translational modification of collagen in normal human articular cartilage. The age-related increase in non-enzymatic glycation affects biomechanical properties of cartilage.* Biochem J, 1998. **330**(Pt 1): p. 345-51.

[152] Hyttinen, M.M., et al., *Age matters: collagen birefringence of superficial articular cartilage is increased in young guinea-pigs but decreased in older animals after identical physiological type of joint loading.* Osteoarthritis Cartilage, 2001. **9**(8): p. 694-701.

[153] Vaughan-Thomas, A., et al., *Modification of the composition of articular cartilage collagen fibrils with increasing age.* Connect Tissue Res, 2008. **49**(5): p. 374-82. DOI: 10.1080/03008200802325417

[154] Axelsson, I. and A. Bjelle, *Proteoglycan structure of bovine articular cartilage. Variation with age and in osteoarthrosis.* Scand J Rheumatol, 1979. **8**(4): p. 217-21. DOI: 10.3109/03009747909114626

[155] Triphaus, G.F., A. Schmidt, and E. Buddecke, *Age-related changes in the incorporation of [35S]sulfate into two proteoglycan populations from human cartilage.* Hoppe Seylers Z Physiol Chem, 1980. **361**(12): p. 1773-9.

[156] Wells, T., et al., *Age-related changes in the composition, the molecular stoichiometry and the stability of proteoglycan aggregates extracted from human articular cartilage.* Biochem J, 2003. **370**(Pt 1): p. 69-79. DOI: 10.1042/BJ20020968

[157] Armstrong, C.G. and V.C. Mow, *Variations in the intrinsic mechanical properties of human articular cartilage with age, degeneration, and water content.* J Bone Joint Surg Am, 1982. **64**(1): p. 88-94.

[158] Ding, M., et al., *Mechanical properties of the normal human tibial cartilage-bone complex in relation to age.* Clin Biomech (Bristol, Avon), 1998. **13**(4-5): p. 351-358.

[159] Athanasiou, K.A., et al., *Effects of aging and dietary restriction on the structural integrity of rat articular cartilage.* Ann Biomed Eng, 2000. **28**(2): p. 143-9. DOI: 10.1114/1.238

[160] Turner, A.S., et al., *Biochemical effects of estrogen on articular cartilage in ovariectomized sheep.* Osteoarthritis Cartilage, 1997. **5**(1): p. 63-9. DOI: 10.1016/S1063-4584(97)80032-5

[161] Athanasiou, K.A., et al., *Effects of diabetes mellitus on the biomechanical properties of human ankle cartilage.* Clin Orthop Relat Res, 1999(368): p. 182-9.

[162] Murray, R.C., et al., *The effects of intra-articular methylprednisolone and exercise on the mechanical properties of articular cartilage in the horse.* Osteoarthritis Cartilage, 1998. **6**(2): p. 106-14. DOI: 10.1053/joca.1997.0100

[163] Cameron, M.L., K.K. Briggs, and J.R. Steadman, *Reproducibility and reliability of the outerbridge classification for grading chondral lesions of the knee arthroscopically.* Am J Sports Med, 2003. **31**(1): p. 83-6.

[164] Outerbridge, R.E., *The etiology of chondromalacia patellae.* J Bone Joint Surg Br, 1961. **43-B**: p. 752-7.

[165] Widuchowski, W., J. Widuchowski, and T. Trzaska, *Articular cartilage defects: study of 25,124 knee arthroscopies.* Knee, 2007. **14**(3): p. 177-82. DOI: 10.1016/j.knee.2007.02.001

[166] Noyes, F.R. and C.L. Stabler, *A system for grading articular cartilage lesions at arthroscopy.* Am J Sports Med, 1989. **17**(4): p. 505-13. DOI: 10.1177/036354658901700410

[167] Beguin, J.A., et al., *[Arthroscopy of the knee. Diagnostic value. 1005 cases]*. Nouv Presse Med, 1982. **11**(49): p. 3619-21.

[168] Acebes, C., et al., *Correlation between arthroscopic and histopathological grading systems of articular cartilage lesions in knee osteoarthritis*. Osteoarthritis Cartilage, 2009. **17**(2): p. 205-12. DOI: 10.1016/j.joca.2008.06.010

[169] Custers, R.J., et al., *Reliability, reproducibility and variability of the traditional Histologic/Histochemical Grading System vs the new OARSI Osteoarthritis Cartilage Histopathology Assessment System*. Osteoarthritis Cartilage, 2007. **15**(11): p. 1241-8. DOI: 10.1016/j.joca.2007.04.017

[170] Radin, E.L., et al., *Effect of repetitive impulsive loading on the knee joints of rabbits*. Clin Orthop, 1978. **131**(131): p. 288-93.

[171] Dekel, S. and S.L. Weissman, *Joint changes after overuse and peak overloading of rabbit knees in vivo*. Acta Orthop Scand, 1978. **49**(6): p. 519-28.

[172] Mankin, H.J., et al., *Form and function of articular cartilage*, in *Orthopaedic basic science*, S. Simon, Editor. 1994, American Academy of Orthopaedic Surgeons: Rosemont, IL.

[173] Buckwalter, J.A., V.C. Mow, and A. Ratcliffe, *Restoration of Injured or Degenerated Articular Cartilage*. J Am Acad Orthop Surg, 1994. **2**(4): p. 192-201.

[174] Furukawa, T., et al., *Biochemical studies on repair cartilage resurfacing experimental defects in the rabbit knee*. J Bone Joint Surg Am, 1980. **62**(1): p. 79-89.

[175] Milentijevic, D. and P.A. Torzilli, *Influence of stress rate on water loss, matrix deformation and chondrocyte viability in impacted articular cartilage*. J Biomech, 2005. **38**(3): p. 493-502. DOI: 10.1016/j.jbiomech.2004.04.016

[176] Komistek, R.D., et al., *Knee mechanics: a review of past and present techniques to determine in vivo loads*. J Biomech, 2005. **38**(2): p. 215-28. DOI: 10.1016/j.jbiomech.2004.02.041

[177] Natoli, R.M. and K.A. Athanasiou, *P188 Reduces Cell Death and IGF-I Reduces GAG Release Following Single-Impact Loading of Articular Cartilage*. J Biomech Eng, 2008. **130**(4): p. 041012. DOI: 10.1115/1.2939368

[178] Wluka, A.E., et al., *Tibial plateau size is related to grade of joint space narrowing and osteophytes in healthy women and in women with osteoarthritis*. Ann Rheum Dis, 2005. **64**(7): p. 1033-7. DOI: 10.1136/ard.2004.029082

[179] Aspden, R.M., J.E. Jeffrey, and L.V. Burgin, *Impact loading of articular cartilage*. Osteoarthritis Cartilage, 2002. **10**(7): p. 588-9; author reply 590.

[180] Anderson, D.D., et al., *A dynamic finite element analysis of impulsive loading of the extension-splinted rabbit knee.* J Biomech Eng, 1990. **112**(2): p. 119-28. DOI: 10.1115/1.2891162

[181] Armstrong, C.G., W.M. Lai, and V.C. Mow, *An analysis of the unconfined compression of articular cartilage.* J Biomech Eng, 1984. **106**(2): p. 165-73. DOI: 10.1115/1.3138475

[182] Atkinson, T.S., R.C. Haut, and N.J. Altiero, *An investigation of biphasic failure criteria for impact-induced fissuring of articular cartilage.* J Biomech Eng, 1998. **120**(4): p. 536-7. DOI: 10.1115/1.2798025

[183] Dunbar, W.L., Jr., et al., *An evaluation of three-dimensional diarthrodial joint contact using penetration data and the finite element method.* J Biomech Eng, 2001. **123**(4): p. 333-40. DOI: 10.1115/1.1384876

[184] Farquhar, T., et al., *Swelling and fibronectin accumulation in articular cartilage explants after cyclical impact.* J Orthop Res, 1996. **14**(3): p. 417-23. DOI: 10.1002/jor.1100140312

[185] Jeffrey, J.E., D.W. Gregory, and R.M. Aspden, *Matrix damage and chondrocyte viability following a single impact load on articular cartilage.* Arch Biochem Biophys, 1995. **322**(1): p. 87-96. DOI: 10.1006/abbi.1995.1439

[186] Jeffrey, J.E., L.A. Thomson, and R.M. Aspden, *Matrix loss and synthesis following a single impact load on articular cartilage in vitro.* Biochim Biophys Acta, 1997. **1334**(2-3): p. 223-32.

[187] D'Lima, D.D., et al., *Human chondrocyte apoptosis in response to mechanical injury.* Osteoarthritis Cartilage, 2001. **9**(8): p. 712-9. DOI: 10.1053/joca.2001.0468

[188] Torzilli, P.A., et al., *Effect of impact load on articular cartilage: cell metabolism and viability, and matrix water content.* J Biomech Eng, 1999. **121**(5): p. 433-41. DOI: 10.1115/1.2835070

[189] Natoli, R.M., C.C. Scott, and K.A. Athanasiou, *Temporal effects of impact on articular cartilage cell death, gene expression, matrix biochemistry, and biomechanics.* Ann Biomed Eng, 2008. **36**(5): p. 780-92. DOI: 10.1007/s10439-008-9472-5

[190] Scott, C.C. and K.A. Athanasiou, *Design, validation, and utilization of an articular cartilage impact instrument.* Proc Inst Mech Eng [H], 2006. **220**(8): p. 845-55.

[191] Blumberg, T.J., R.M. Natoli, and K.A. Athanasiou, *Effects of doxycycline on articular cartilage GAG release and mechanical properties following impact.* Biotechnol Bioeng, 2008. **100**(3): p. 506-15. DOI: 10.1002/bit.21778

[192] Donohue, J.M., et al., *The effects of indirect blunt trauma on adult canine articular cartilage.* J Bone Joint Surg Am, 1983. **65**(7): p. 948-57.

[193] Thompson, R.C., Jr. and C.A. Bassett, *Histological observations on experimentally induced degeneration of articular cartilage.* J Bone Joint Surg Am, 1970. **52**(3): p. 435-43.

[194] Simon, S.R., et al., *The response of joints to impact loading. II. In vivo behavior of subchondral bone.* J Biomech, 1972. **5**(3): p. 267-72. DOI: 10.1016/0021-9290(72)90042-5

[195] Mitchell, N. and N. Shepard, *Healing of articular cartilage in intra-articular fractures in rabbits.* J Bone Joint Surg Am, 1980. **62**(4): p. 628-34.

[196] Haut, R.C., T.M. Ide, and C.E. De Camp, *Mechanical responses of the rabbit patello-femoral joint to blunt impact.* J Biomech Eng, 1995. **117**(4): p. 402-8. DOI: 10.1115/1.2794199

[197] Ewers, B.J., et al., *Rate of blunt impact loading affects changes in retropatellar cartilage and underlying bone in the rabbit patella.* J Biomech, 2002. **35**(6): p. 747-55. DOI: 10.1016/S0021-9290(02)00019-2

[198] Ewers, B.J., et al., *Chronic changes in rabbit retro-patellar cartilage and subchondral bone after blunt impact loading of the patellofemoral joint.* J Orthop Res, 2002. **20**(3): p. 545-50. DOI: 10.1016/S0736-0266(01)00135-8

[199] Bonassar, L.J., et al., *The effect of dynamic compression on the response of articular cartilage to insulin-like growth factor-I.* J Orthop Res, 2001. **19**(1): p. 11-7. DOI: 10.1016/S0736-0266(00)00004-8

[200] Morel, V. and T.M. Quinn, *Cartilage injury by ramp compression near the gel diffusion rate.* J Orthop Res, 2004. **22**(1): p. 145-51. DOI: 10.1016/S0736-0266(03)00164-5

[201] Webb, J., *Crystal induced arthritis: gout and pseudogout.* Aust Fam Physician, 1978. **7**(8): p. 959-80.

[202] Kozin, F. and D.J. McCarty, *Protein adsorption to monosodium urate, calcium pyrophosphate dihydrate, and silica crystals: relationship to the pathogenesis of crystal-induced inflammation.* Arthritis Rheum, 1976. **19**Suppl 3: p. 433-8. DOI: 10.1002/1529-0131(197605/06)19:3+<433::AID-ART1780190718>3.0.CO;2-U

[203] Mandel, N.S. and G.S. Mandel, *Monosodium urate monohydrate, the gout culprit.* J Am Chem Soc, 1976. **98**(8): p. 2319-23. DOI: 10.1021/ja00424a054

[204] Muehleman, C., et al., *Association between crystals and cartilage degeneration in the ankle.* J Rheumatol, 2008. **35**(6): p. 1108-17.

[205] Rizzoli, A.J., L. Trujeque, and A.D. Bankhurst, *The coexistence of gout and rheumatoid arthritis: case reports and a review of the literature.* J Rheumatol, 1980. **7**(3): p. 316-24.

[206] Shields, K.J., J.R. Owen, and J.S. Wayne, *Biomechanical and biotribological correlation of induced wear on bovine femoral condyles.* J Biomech Eng, 2009. **131**(6): p. 061005. DOI: 10.1115/1.3116156

[207] Cheung, H.S., et al., *In vitro collagen biosynthesis in healing and normal rabbit articular cartilage.* J Bone Joint Surg Am, 1978. **60**(8): p. 1076-81.

[208] DePalma, A.F., C.D. McKeever, and D.K. Subin, *Process of repair of articular cartilage demonstrated by histology and autoradiography with tritiated thymidine.* Clin Orthop Relat Res, 1966. **48**: p. 229-42. DOI: 10.1097/00003086-196609000-00028

[209] Fuller, J.A. and F.N. Ghadially, *Ultrastructural observations on surgically produced partial-thickness defects in articular cartilage.* Clin Orthop Relat Res, 1972. **86**: p. 193-205. DOI: 10.1097/00003086-197207000-00031

[210] Shapiro, F., S. Koide, and M.J. Glimcher, *Cell origin and differentiation in the repair of full-thickness defects of articular cartilage.* J Bone Joint Surg Am, 1993. **75**(4): p. 532-53.

[211] Hjertquist, S.O. and R. Lemperg, *Histological, autoradiographic and microchemical studies of spontaneously healing osteochondral articular defects in adult rabbits.* Calcif Tissue Res, 1971. **8**(1): p. 54-72. DOI: 10.1007/BF02010122

[212] Cheung, H.S., et al., *In vitro synthesis of tissue-specific type II collagen by healing cartilage. I. Short-term repair of cartilage by mature rabbits.* Arthritis Rheum, 1980. **23**(2): p. 211-9. DOI: 10.1002/art.1780230212

[213] Ghadially, J.A., R. Ghadially, and F.N. Ghadially, *Long-term results of deep defects in articular cartilage. A scanning electron microscope study.* Virchows Arch B Cell Pathol, 1977. **25**(2): p. 125-36.

[214] Squires, G.R., et al., *The pathobiology of focal lesion development in aging human articular cartilage and molecular matrix changes characteristic of osteoarthritis.* Arthritis Rheum, 2003. **48**(5): p. 1261-70. DOI: 10.1002/art.10976

[215] Administration, U.S.F.a.D., *Cellular products for joint surface repair - Briefing document.* Meeting 38 of the Cellular, tissue, and gene therapies advisory committee. 2005.

[216] Upmeier, H., et al., *Follow-up costs up to 5 years after conventional treatments in patients with cartilage lesions of the knee.* Knee Surg Sports Traumatol Arthrosc, 2007. **15**(3): p. 249-57. DOI: 10.1007/s00167-006-0182-y

[217] Masini, B.D., et al., *Resource utilization and disability outcome assessment of combat casualties from Operation Iraqi Freedom and Operation Enduring Freedom.* J Orthop Trauma, 2009. **23**(4): p. 261-6. DOI: 10.1097/BOT.0b013e31819dfa04

[218] Jackson, D.W., T.M. Simon, and H.M. Aberman, *Symptomatic articular cartilage degeneration: the impact in the new millennium.* Clin Orthop, 2001. **391**(Suppl): p. S14-25.

[219] Cunningham, L.S. and J.L. Kelsey, *Epidemiology of musculoskeletal impairments and associated disability.* Am J Public Health, 1984. **74**(6): p. 574-9. DOI: 10.2105/AJPH.74.6.574

[220] Lawrence, R.C., et al., *Estimates of the prevalence of arthritis and other rheumatic conditions in the United States. Part II.* Arthritis Rheum, 2008. **58**(1): p. 26-35. DOI: 10.1002/art.23176

[221] Hart, D.J., D.V. Doyle, and T.D. Spector, *Incidence and risk factors for radiographic knee osteoarthritis in middle-aged women: the Chingford Study.* Arthritis Rheum, 1999. **42**(1): p. 17-24. DOI: 10.1002/1529-0131(199901)42:1<17::AID-ANR2>3.0.CO;2-E

[222] Buckwalter, J.A., *Articular cartilage injuries.* Clin Orthop Relat Res, 2002. **402**(402): p. 21-37.

[223] Lohmander, L.S., et al., *High prevalence of knee osteoarthritis, pain, and functional limitations in female soccer players twelve years after anterior cruciate ligament injury.* Arthritis Rheum, 2004. **50**(10): p. 3145-52. DOI: 10.1002/art.20589

[224] Roos, H., et al., *Osteoarthritis of the knee after injury to the anterior cruciate ligament or meniscus: the influence of time and age.* Osteoarthritis Cartilage, 1995. **3**(4): p. 261-7. DOI: 10.1016/S1063-4584(05)80017-2

[225] Borrelli, J., Jr. and W.M. Ricci, *Acute effects of cartilage impact.* Clin Orthop, 2004. **423**: p. 33-9.

[226] Chrisman, O.D., et al., *1981 Nicolas Andry Award. The relationship of mechanical trauma and the early biochemical reactions of osteoarthritic cartilage.* Clin Orthop, 1981. **161**(161): p. 275-84.

[227] Olson, S.A. and F. Guilak, *From articular fracture to posttraumatic arthritis: a black box that needs to be opened.* J Orthop Trauma, 2006. **20**(10): p. 661-2. DOI: 10.1097/01.bot.0000245683.89152.55

[228] Gabriel, S.E. and K. Michaud, *Epidemiological studies in incidence, prevalence, mortality, and comorbidity of the rheumatic diseases.* Arthritis Res Ther, 2009. **11**(3): p. 229. DOI: 10.1186/ar2669

[229] Eskelinen, A.P., et al., *Primary cartilage lesions of the knee joint in young male adults. Overweight as a predisposing factor. An arthroscopic study.* Scand J Surg, 2004. **93**(3): p. 229-33.

[230] Butler, R.J., et al., *Gait mechanics after ACL reconstruction: implications for the early onset of knee osteoarthritis.* Br J Sports Med, 2009. **43**(5): p. 366-70. DOI: 10.1136/bjsm.2008.052522

[231] Mills, P.M., et al., *Tibio-femoral cartilage defects 3-5 years following arthroscopic partial medial meniscectomy.* Osteoarthritis Cartilage, 2008. **16**(12): p. 1526-31. DOI: 10.1016/j.joca.2008.04.014

[232] Appleyard, R.C., P. Ghosh, and M.V. Swain, *Biomechanical, histological and immunohistological studies of patellar cartilage in an ovine model of osteoarthritis induced by lateral meniscectomy.* Osteoarthritis Cartilage, 1999. **7**(3): p. 281-94. DOI: 10.1053/joca.1998.0202

[233] Roos, H., et al., *Knee osteoarthritis after meniscectomy: prevalence of radiographic changes after twenty-one years, compared with matched controls.* Arthritis Rheum, 1998. **41**(4): p. 687-93. DOI: 10.1002/1529-0131(199804)41:4<687::AID-ART16>3.0.CO;2-2

[234] Hunter, W., *Of the structure and diseases of articulating cartilages.* Philos Trans R Soc, 1743. **43**: p. 514-521.

[235] Kern, D., M.B. Zlatkin, and M.K. Dalinka, *Occupational and post-traumatic arthritis.* Radiol Clin North Am, 1988. **26**(6): p. 1349-58.

[236] Andriacchi, T.P. and C.O. Dyrby, *Interactions between kinematics and loading during walking for the normal and ACL deficient knee.* J Biomech, 2005. **38**(2): p. 293-8. DOI: 10.1016/j.jbiomech.2004.02.010

[237] Georgoulis, A.D., et al., *Three-dimensional tibiofemoral kinematics of the anterior cruciate ligament-deficient and reconstructed knee during walking.* Am J Sports Med, 2003. **31**(1): p. 75-9.

[238] Li, G., et al., *Anterior cruciate ligament deficiency alters the in vivo motion of the tibiofemoral cartilage contact points in both the anteroposterior and mediolateral directions.* J Bone Joint Surg Am, 2006. **88**(8): p. 1826-34. DOI: 10.2106/JBJS.E.00539

[239] Andriacchi, T.P., et al., *Rotational changes at the knee after ACL injury cause cartilage thinning.* Clin Orthop Relat Res, 2006. **442**: p. 39-44. DOI: 10.1097/01.blo.0000197079.26600.09

[240] Andriacchi, T.P., et al., *A framework for the in vivo pathomechanics of osteoarthritis at the knee.* Ann Biomed Eng, 2004. **32**(3): p. 447-57. DOI: 10.1023/B:ABME.0000017541.82498.37

[241] Hurley, M.V., D.W. Jones, and D.J. Newham, *Arthrogenic quadriceps inhibition and rehabilitation of patients with extensive traumatic knee injuries.* Clin Sci (Lond), 1994. **86**(3): p. 305-10.

[242] McVey, E.D., et al., *Arthrogenic muscle inhibition in the leg muscles of subjects exhibiting functional ankle instability.* Foot Ankle Int, 2005. **26**(12): p. 1055-61.

[243] Palmieri, R.M., et al., *Arthrogenic muscle response to a simulated ankle joint effusion.* Br J Sports Med, 2004. **38**(1): p. 26-30. DOI: 10.1136/bjsm.2002.001677

[244] Sedory, E.J., et al., *Arthrogenic muscle response of the quadriceps and hamstrings with chronic ankle instability.* J Athl Train, 2007. **42**(3): p. 355-60.

[245] Palmieri-Smith, R.M. and A.C. Thomas, *A Neuromuscular mechanism of posttraumatic osteoarthritis associated with ACL injury.* Exerc Sport Sci Rev, 2009. **37**(3): p. 147-53. DOI: 10.1097/JES.0b013e3181aa6669

[246] Johnson, R.G. and A.R. Poole, *The early response of articular cartilage to ACL transection in a canine model.* Exp Pathol, 1990. **38**(1): p. 37-52.

[247] Pond, M.J. and G. Nuki, *Experimentally-induced osteoarthritis in the dog.* Ann Rheum Dis, 1973. **32**(4): p. 387-8. DOI: 10.1136/ard.32.4.387

[248] Guilak, F., et al., *The effects of matrix compression on proteoglycan metabolism in articular cartilage explants.* Osteoarthritis Cartilage, 1994. **2**(2): p. 91-101. DOI: 10.1016/S1063-4584(05)80059-7

[249] Kisiday, J.D., et al., *Catabolic responses of chondrocyte-seeded peptide hydrogel to dynamic compression.* Ann Biomed Eng, 2009. **37**(7): p. 1368-75. DOI: 10.1007/s10439-009-9699-9

[250] Ragan, P.M., et al., *Chondrocyte extracellular matrix synthesis and turnover are influenced by static compression in a new alginate disk culture system.* Arch Biochem Biophys, 2000. **383**(2): p. 256-64. DOI: 10.1006/abbi.2000.2060

[251] Healy, Z.R., et al., *Divergent responses of chondrocytes and endothelial cells to shear stress: cross-talk among COX-2, the phase 2 response, and apoptosis.* Proc Natl Acad Sci U S A, 2005. **102**(39): p. 14010-5. DOI: 10.1073/pnas.0506620102

[252] Smith, R.L., et al., *Effects of fluid-induced shear on articular chondrocyte morphology and metabolism in vitro.* J Orthop Res, 1995. **13**(6): p. 824-31. DOI: 10.1002/jor.1100130604

[253] Kuo, C.H. and L.M. Keer, *Contact Stress and Fracture Analysis of Articular Cartilage.* Bimoed Eng Appl Basis Comm, 1993. **5**: p. 515-521.

[254] Wu, J.Z., W. Herzog, and J. Ronsky, *Modeling axi-symmetrical joint contact with biphasic cartilage layers–an asymptotic solution.* J Biomech, 1996. **29**(10): p. 1263-81. DOI: 10.1016/0021-9290(96)00051-6

[255] Kafka, V., *Surface fissures in articular cartilage: new concepts, hypotheses and modeling.* Clin Biomech (Bristol, Avon), 2002. **17**(1): p. 73-80.

[256] Wilson, W., et al., *Causes of mechanically induced collagen damage in articular cartilage.* J Orthop Res, 2006. **24**(2): p. 220-8. DOI: 10.1002/jor.20027

[257] Baars, D.C., S.A. Rundell, and R.C. Haut, *Treatment with the non-ionic surfactant poloxamer P188 reduces DNA fragmentation in cells from bovine chondral explants exposed to injurious unconfined compression.* Biomech Model Mechanobiol, 2006. **5**(2-3): p. 133-9. DOI: 10.1007/s10237-006-0024-3

[258] Phillips, D.M. and R.C. Haut, *The use of a non-ionic surfactant (P188) to save chondrocytes from necrosis following impact loading of chondral explants.* J Orthop Res, 2004. **22**(5): p. 1135-42. DOI: 10.1016/j.orthres.2004.02.002

[259] Brandt, K.D., et al., *Osteoarthritic changes in canine articular cartilage, subchondral bone, and synovium fifty-four months after transection of the anterior cruciate ligament.* Arthritis Rheum, 1991. **34**(12): p. 1560-70.

[260] Egan, M.S., et al., *The association of amyloid deposits and osteoarthritis.* Arthritis Rheum, 1982. **25**(2): p. 204-8. DOI: 10.1002/art.1780250214

[261] Sorensen, K.H. and H.E. Christensen, *Local amyloid formation in the hip joint capsule in osteoarthritis.* Acta Orthop Scand, 1973. **44**(4): p. 460-6.

[262] Kusakabe, A., *Subchondral cancellous bone in osteoarthrosis and rheumatoid arthritis of the femoral head. A quantitative histological study of trabecular remodelling.* Arch Orthop Unfallchir, 1977. **88**(2): p. 185-97. DOI: 10.1007/BF00415099

[263] Oettmeier, R. and K. Abendroth, *Osteoarthritis and bone: osteologic types of osteoarthritis of the hip.* Skeletal Radiol, 1989. **18**(3): p. 165-74. DOI: 10.1007/BF00360962

[264] van der Esch, M., et al., *Structural joint changes, malalignment, and laxity in osteoarthritis of the knee.* Scand J Rheumatol, 2005. **34**(4): p. 298-301.

[265] Felson, D.T., et al., *Osteophytes and progression of knee osteoarthritis.* Rheumatology (Oxford), 2005. **44**(1): p. 100-4. DOI: 10.1093/rheumatology/keh411

[266] Nagaosa, Y., P. Lanyon, and M. Doherty, *Characterisation of size and direction of osteophyte in knee osteoarthritis: a radiographic study.* Ann Rheum Dis, 2002. **61**(4): p. 319-24. DOI: 10.1136/ard.61.4.319

[267] Poole, C.A., *Articular cartilage chondrons: form, function and failure.* J Anat, 1997. **191**(Pt 1): p. 1-13. DOI: 10.1046/j.1469-7580.1997.19110001.x

[268] Polzer, K., G. Schett, and J. Zwerina, *The lonely death: chondrocyte apoptosis in TNF-induced arthritis.* Autoimmunity, 2007. **40**(4): p. 333-6. DOI: 10.1080/08916930701356721

[269] Attur, M.G., et al., *F-spondin, a neuroregulatory protein, is up-regulated in osteoarthritis and regulates cartilage metabolism via TGF-beta activation.* FASEB J, 2009. **23**(1): p. 79-89. DOI: 10.1096/fj.08-114363

[270] Recklies, A.D., L. Baillargeon, and C. White, *Regulation of cartilage oligomeric matrix protein synthesis in human synovial cells and articular chondrocytes.* Arthritis Rheum, 1998. **41**(6): p. 997-1006. DOI: 10.1002/1529-0131(199806)41:6<997::AID-ART6>3.0.CO;2-G

[271] Yelin, E., *The economics of osteoarthritis,* in *Osteoarthritis,* K.D. Brandt, M. Doherty, and L.S. Lohmander, Editors. 1998, Oxford University Press: New York. p. 23-30.

[272] Lawrence, R.C., et al., *Estimates of the prevalence of arthritis and selected musculoskeletal disorders in the United States.* Arthritis Rheum, 1998. **41**(5): p. 778-99. DOI: 10.1002/1529-0131(199805)41:5<778::AID-ART4>3.0.CO;2-V

[273] Gabriel, S.E., et al., *Direct medical costs unique to people with arthritis.* J Rheumatol, 1997. **24**(4): p. 719-25.

[274] Shea, K.G., P.J. Apel, and R.P. Pfeiffer, *Anterior cruciate ligament injury in paediatric and adolescent patients: a review of basic science and clinical research.* Sports Med, 2003. **33**(6): p. 455-71. DOI: 10.2165/00007256-200333060-00006

[275] Utukuri, M.M., et al., *Update on paediatric ACL injuries.* Knee, 2006. **13**(5): p. 345-52. DOI: 10.1016/j.knee.2006.06.001

[276] Adirim, T.A. and T.L. Cheng, *Overview of injuries in the young athlete.* Sports Med, 2003. **33**(1): p. 75-81. DOI: 10.2165/00007256-200333010-00006

[277] Berler, R., *Arms-Control Breakdown*, in *NY Times*. 2009, Arthur Ochs Sulzberger, Jr.: New York. p. MM20.

[278] Louw, Q.A., J. Manilall, and K.A. Grimmer, *Epidemiology of knee injuries among adolescents: a systematic review.* Br J Sports Med, 2008. **42**(1): p. 2-10.

[279] Jones, D., Q. Louw, and K. Grimmer, *Recreational and sporting injury to the adolescent knee and ankle: Prevalence and causes.* Aust J Physiother, 2000. **46**(3): p. 179-188.

[280] Hewett, T.E., et al., *Biomechanical measures of neuromuscular control and valgus loading of the knee predict anterior cruciate ligament injury risk in female athletes: a prospective study.* Am J Sports Med, 2005. **33**(4): p. 492-501. DOI: 10.1177/0363546504269591

[281] Gomez, E., J.C. DeLee, and W.C. Farney, *Incidence of injury in Texas girls' high school basketball.* Am J Sports Med, 1996. **24**(5): p. 684-7. DOI: 10.1177/036354659602400521

[282] Hickey, G.J., P.A. Fricker, and W.A. McDonald, *Injuries of young elite female basketball players over a six-year period.* Clin J Sport Med, 1997. **7**(4): p. 252-6. DOI: 10.1097/00042752-199710000-00002

[283] Powell, J.W. and K.D. Barber-Foss, *Injury Patterns in Selected High School Sports: A Review of the 1995-1997 Seasons.* J Athl Train, 1999. **34**(3): p. 277-284.

[284] Hunter, D.J., et al., *Structural factors associated with malalignment in knee osteoarthritis: the Boston osteoarthritis knee study.* J Rheumatol, 2005. **32**(11): p. 2192-9.

[285] Sharma, L., et al., *Relationship of meniscal damage, meniscal extrusion, malalignment, and joint laxity to subsequent cartilage loss in osteoarthritic knees.* Arthritis Rheum, 2008. **58**(6): p. 1716-26. DOI: 10.1002/art.23462

[286] Mikos, A.G., et al., *Engineering complex tissues.* Tissue Eng, 2006. **12**(12): p. 3307-39. DOI: 10.1089/ten.2006.12.3307

[287] *United States Bone and Joint Decade: The Burden of Musculoskeletal Diseases in the United States.* 2008, Rosemont, IL: American Academy of Orthopaedic Surgeons.

[288] Curl, W.W., et al., *Cartilage injuries: a review of 31,516 knee arthroscopies.* Arthroscopy, 1997. **13**(4): p. 456-60.

[289] Chesterman, P.J. and A.U. Smith, *Homotransplantation of articular cartilage and isolated chondrocytes. An experimental study in rabbits.* J Bone Joint Surg Br, 1968. **50**(1): p. 184-97.

[290] Hunziker, E.B., *Articular cartilage repair: basic science and clinical progress. A review of the current status and prospects.* Osteoarthritis Cartilage, 2002. **10**(6): p. 432-63. DOI: 10.1053/joca.2002.0801

[291] Smith, G.D., G. Knutsen, and J.B. Richardson, *A clinical review of cartilage repair techniques.* J Bone Joint Surg Br, 2005. **87**(4): p. 445-9.

[292] Moutos, F.T., L.E. Freed, and F. Guilak, *A biomimetic three-dimensional woven composite scaffold for functional tissue engineering of cartilage.* Nat Mater, 2007. **6**(2): p. 162-7. DOI: 10.1038/nmat1822

[293] Vanwanseele, B., et al., *Knee cartilage of spinal cord-injured patients displays progressive thinning in the absence of normal joint loading and movement.* Arthritis Rheum, 2002. **46**(8): p. 2073-8. DOI: 10.1002/art.10462

[294] Guilak, F., *Functional tissue engineering: the role of biomechanics in reparative medicine.* Annals of the New York Academy of Sciences, 2002. **961**: p. 193-5.

[295] Tatebe, M., et al., *Differentiation of transplanted mesenchymal stem cells in a large osteochondral defect in rabbit.* Cytotherapy, 2005. **7**(6): p. 520-30. DOI: 10.1080/14653240500361350

[296] Homminga, G.N., et al., *Repair of sheep articular cartilage defects with a rabbit costal perichondrial graft.* Acta Orthop Scand, 1991. **62**(5): p. 415-8.

[297] van Susante, J.L., et al., *Resurfacing potential of heterologous chondrocytes suspended in fibrin glue in large full-thickness defects of femoral articular cartilage: an experimental study in the goat.* Biomaterials, 1999. **20**(13): p. 1167-75.

[298] Toolan, B.C., et al., *Development of a novel osteochondral graft for cartilage repair.* J Biomed Mater Res, 1998. **41**(2): p. 244-50. DOI: 10.1002/(SICI)1097-4636(199808)41:2<244::AID-JBM9>3.0.CO;2-I

[299] Hwang, N.S. and J. Elisseeff, *Application of stem cells for articular cartilage regeneration.* J Knee Surg, 2009. **22**(1): p. 60-71.

[300] Chen, F.H. and R.S. Tuan, *Mesenchymal stem cells in arthritic diseases.* Arthritis Res Ther, 2008. **10**(5): p. 223. DOI: 10.1186/ar2514

[301] Deng, Y., J.C. Hu, and K.A. Athanasiou, *Isolation and chondroinduction of a dermis-isolated, aggrecan-sensitive subpopulation with high chondrogenic potential.* Arthritis Rheum, 2007. **56**(1): p. 168-76. DOI: 10.1002/art.22300

[302] Banfi, A., et al., *Replicative aging and gene expression in long-term cultures of human bone marrow stromal cells.* Tissue Eng, 2002. **8**(6): p. 901-10. DOI: 10.1089/107632702320934001

[303] Baxter, M.A., et al., *Study of telomere length reveals rapid aging of human marrow stromal cells following in vitro expansion.* Stem Cells, 2004. **22**(5): p. 675-82. DOI: 10.1634/stemcells.22-5-675

[304] Bruder, S.P., N. Jaiswal, and S.E. Haynesworth, *Growth kinetics, self-renewal, and the osteogenic potential of purified human mesenchymal stem cells during extensive sub-cultivation and following cryopreservation.* J Cell Biochem, 1997. **64**(2): p. 278-94. DOI: 10.1002/(SICI)1097-4644(199702)64:2<278::AID-JCB11>3.0.CO;2-F

[305] Parsch, D., et al., *Telomere length and telomerase activity during expansion and differentiation of human mesenchymal stem cells and chondrocytes.* J Mol Med, 2004. **82**(1): p. 49-55.

[306] Vacanti, V., et al., *Phenotypic changes of adult porcine mesenchymal stem cells induced by prolonged passaging in culture.* J Cell Physiol, 2005. **205**(2): p. 194-201. DOI: 10.1002/jcp.20376

[307] Koay, E.J., G.M. Hoben, and K.A. Athanasiou, *Tissue engineering with chondrogenically differentiated human embryonic stem cells.* Stem Cells, 2007. **25**(9): p. 2183-90. DOI: 10.1634/stemcells.2007-0105

[308] Koay, E.J. and K.A. Athanasiou, *Development of serum-free, chemically defined conditions for human embryonic stem cell-derived fibrochondrogenesis.* Tissue Eng Part A, 2009. **15**(8): p. 2249-57. DOI: 10.1089/ten.tea.2008.0320

[309] Koay, E.J. and K.A. Athanasiou, *Hypoxic chondrogenic differentiation of human embryonic stem cells enhances cartilage protein synthesis and biomechanical functionality.* Osteoarthritis Cartilage, 2008. **16**(12): p. 1450-6. DOI: 10.1016/j.joca.2008.04.007

[310] Hoben, G.M., E.J. Koay, and K.A. Athanasiou, *Fibrochondrogenesis in two embryonic stem cell lines: effects of differentiation timelines.* Stem Cells, 2008. **26**(2): p. 422-30. DOI: 10.1634/stemcells.2007-0641

[311] Jubel, A., et al., *Transplantation of de novo scaffold-free cartilage implants into sheep knee chondral defects.* Am J Sports Med, 2008. **36**(8): p. 1555-64. DOI: 10.1177/0363546508321474

[312] Hu, J.C. and K.A. Athanasiou, *The effects of intermittent hydrostatic pressure on self-assembled articular cartilage constructs.* Tissue Eng, 2006. **12**(5): p. 1337-44. DOI: 10.1089/ten.2006.12.1337

[313] Hu, J.C. and K.A. Athanasiou, *A self-assembling process in articular cartilage tissue engineering.* Tissue Eng, 2006. **12**(4): p. 969-79. DOI: 10.1089/ten.2006.12.969

[314] Ofek, G., et al., *Matrix development in self-assembly of articular cartilage.* PLoS ONE, 2008. **3**(7): p. e2795. DOI: 10.1371/journal.pone.0002795

[315] Dobratz, E.J., et al., *Injectable cartilage: using alginate and human chondrocytes.* Arch Facial Plast Surg, 2009. **11**(1): p. 40-7. DOI: 10.1001/archfacial.2008.509

[316] Park, K.M., et al., *Thermosensitive chitosan-Pluronic hydrogel as an injectable cell delivery carrier for cartilage regeneration.* Acta Biomater, 2009. DOI: 10.1016/j.actbio.2009.01.040

[317] Bryant, S.J. and K.S. Anseth, *The effects of scaffold thickness on tissue engineered cartilage in photocrosslinked poly(ethylene oxide) hydrogels.* Biomaterials, 2001. **22**(6): p. 619-26. DOI: 10.1016/S0142-9612(00)00225-8

[318] Murphy, C.L. and A. Sambanis, *Effect of oxygen tension and alginate encapsulation on restoration of the differentiated phenotype of passaged chondrocytes.* Tissue Eng, 2001. **7**(6): p. 791-803. DOI: 10.1089/107632701753337735

[319] Hutmacher, D.W., *Scaffold design and fabrication technologies for engineering tissues - state of the art and future perspectives.* J Biomater Sci Polym Ed, 2001. **12**(1): p. 107-24. DOI: 10.1163/156856201744489

[320] Saris, D.B., et al., *Dynamic pressure transmission through agarose gels.* Tissue Eng, 2000. **6**(5): p. 531-7. DOI: 10.1089/107632700750022170

[321] Elder, S.H., et al., *Chondrocyte differentiation is modulated by frequency and duration of cyclic compressive loading.* Ann Biomed Eng, 2001. **29**(6): p. 476-82. DOI: 10.1114/1.1376696

[322] Mauck, R.L., et al., *Functional tissue engineering of articular cartilage through dynamic loading of chondrocyte-seeded agarose gels.* J Biomech Eng, 2000. **122**(3): p. 252-60. DOI: 10.1115/1.429656

[323] Hunter, C.J., et al., *Mechanical compression alters gene expression and extracellular matrix synthesis by chondrocytes cultured in collagen I gels.* Biomaterials, 2002. **23**(4): p. 1249-59. DOI: 10.1016/S0142-9612(01)00245-9

[324] Lee, C.R., et al., *Articular cartilage chondrocytes in type I and type II collagen-GAG matrices exhibit contractile behavior in vitro.* Tissue Eng, 2000. **6**(5): p. 555-65. DOI: 10.1089/107632700750022198

[325] van Susante, J.L.C., et al., *Linkage of chondroitin-sulfate to type I collagen scaffolds stimulates the bioactivity of seeded chondrocytes in vitro.* Biomaterials, 2001. **22**(17): p. 2359-69.

[326] Nehrer, S., et al., *Canine chondrocytes seeded in type I and type II collagen implants investigated in vitro [published erratum appears in J Biomed Mater Res 1997 Winter;38(4):288].* J Biomed Mater Res, 1997. **38**(2): p. 95-104.

[327] Madihally, S.V. and H.W. Matthew, *Porous chitosan scaffolds for tissue engineering.* Biomaterials, 1999. **20**(12): p. 1133-42. DOI: 10.1016/S0142-9612(99)00011-3

[328] Sechriest, V.F., et al., *GAG-augmented polysaccharide hydrogel: a novel biocompatible and biodegradable material to support chondrogenesis.* J Biomed Mater Res, 2000. **49**(4): p. 534-41. DOI: 10.1002/(SICI)1097-4636(20000315)49:4<534::AID-JBM12>3.0.CO;2-#

[329] Chupa, J.M., et al., *Vascular cell responses to polysaccharide materials: in vitro and in vivo evaluations.* Biomaterials, 2000. **21**(22): p. 2315-22. DOI: 10.1016/S0142-9612(00)00158-7

[330] Wang, Y., et al., *Stem cell-based tissue engineering with silk biomaterials.* Biomaterials, 2006. **27**(36): p. 6064-82. DOI: 10.1016/j.biomaterials.2006.07.008

[331] Meinel, L., et al., *Engineering cartilage-like tissue using human mesenchymal stem cells and silk protein scaffolds.* Biotechnol Bioeng, 2004. **88**(3): p. 379-91. DOI: 10.1002/bit.20252

[332] Silverman, R.P., et al., *Injectable tissue-engineered cartilage using a fibrin glue polymer.* Plast Reconstr Surg, 1999. **103**(7): p. 1809-18.

[333] Almqvist, K.F., et al., *Culture of chondrocytes in alginate surrounded by fibrin gel: characteristics of the cells over a period of eight weeks.* Ann Rheum Dis, 2001. **60**(8): p. 781-90. DOI: 10.1136/ard.60.8.781

[334] Perka, C., et al., *The use of fibrin beads for tissue engineering and subsequential transplantation.* Tissue Eng, 2001. **7**(3): p. 359-61. DOI: 10.1089/10763270152044215

[335] Perka, C., et al., *Joint cartilage repair with transplantation of embryonic chondrocytes embedded in collagen-fibrin matrices.* Clin Exp Rheumatol, 2000. **18**(1): p. 13-22.

[336] Dare, E.V., et al., *Genipin Cross-Linked Fibrin Hydrogels for in vitro Human Articular Cartilage Tissue-Engineered Regeneration.* Cells Tissues Organs, 2009. DOI: 10.1159/000209230

[337] Grigolo, B., et al., *Transplantation of chondrocytes seeded on a hyaluronan derivative (hyaff-11) into cartilage defects in rabbits.* Biomaterials, 2001. **22**(17): p. 2417-24. DOI: 10.1016/S0142-9612(00)00429-4

[338] Solchaga, L.A., et al., *Hyaluronan-based polymers in the treatment of osteochondral defects.* J Orthop Res, 2000. **18**(5): p. 773-80. DOI: 10.1002/jor.1100180515

[339] Badylak, S.F., *The extracellular matrix as a biologic scaffold material.* Biomaterials, 2007. **28**(25): p. 3587-93. DOI: 10.1016/j.biomaterials.2007.04.043

[340] Urist, M.R., *Bone: formation by autoinduction*. Science, 1965. **150**(698): p. 893-9. DOI: 10.1126/science.150.3698.893

[341] Ott, H.C., et al., *Perfusion-decellularized matrix: using nature's platform to engineer a bioartificial heart*. Nat Med, 2008. **14**(2): p. 213-21. DOI: 10.1038/nm1684

[342] Macchiarini, P., et al., *Clinical transplantation of a tissue-engineered airway*. Lancet, 2008. **372**(9655): p. 2023-30.

[343] Cheng, N.C., et al., *Chondrogenic Differentiation of Adipose-Derived Adult Stem Cells by a Porous Scaffold Derived from Native Articular Cartilage Extracellular Matrix*. Tissue Eng Part A, 2008.

[344] Kleinman, H.K., et al., *Isolation and characterization of type IV procollagen, laminin, and heparan sulfate proteoglycan from the EHS sarcoma*. Biochemistry, 1982. **21**(24): p. 6188-93. DOI: 10.1021/bi00267a025

[345] Athanasiou, K.A., et al., *Orthopaedic applications for PLA-PGA biodegradable polymers*. Arthroscopy, 1998. **14**(7): p. 726-37.

[346] Athanasiou, K.A., D. Korvick, and R.C. Schenck, *Biodegradable implants for the treatment of osteochondral defects in a goat model*. Tissue Eng, 1997. **3**(4): p. 363-373. DOI: 10.1089/ten.1997.3.363

[347] Athanasiou, K.A., G.G. Niederauer, and C.M. Agrawal, *Sterilization, toxicity, biocompatibility and clinical applications of polylactic acid/polyglycolic acid copolymers*. Biomaterials, 1996. **17**(2): p. 93-102. DOI: 10.1016/0142-9612(96)85754-1

[348] Athanasiou, K.A., et al., *Applications of biodegradable lactides and glycolides in podiatry*. Clin Podiatr Med Surg, 1995. **12**(3): p. 475-95.

[349] Athanasiou, K.A., et al., *In vitro degradation and release characteristics of biodegradable implants containing trypsin inhibitor*. Clin Orthop, 1995(315): p. 272-81.

[350] Vunjak-Novakovic, G. and L.E. Freed, *Culture of organized cell communities*. Adv Drug Deliv Rev, 1998. **33**(1-2): p. 15-30.

[351] Daniels, A.U., M.K. Chang, and K.P. Andriano, *Mechanical properties of biodegradable polymers and composites proposed for internal fixation of bone*. J Appl Biomater, 1990. **1**(1): p. 57-78. DOI: 10.1002/jab.770010109

[352] Freed, L.E., et al., *Joint resurfacing using allograft chondrocytes and synthetic biodegradable polymer scaffolds*. J Biomed Mater Res, 1994. **28**(8): p. 891-9. DOI: 10.1002/jbm.820280808

[353] Freed, L.E., et al., *Chondrogenesis in a cell-polymer-bioreactor system*. Exp Cell Res, 1998. **240**(1): p. 58-65. DOI: 10.1006/excr.1998.4010

[354] Freed, L.E., G. Vunjak-Novakovic, and R. Langer, *Cultivation of cell-polymer cartilage implants in bioreactors.* J Cell Biochem, 1993. **51**(3): p. 257-64. DOI: 10.1002/jcb.240510304

[355] Grande, D.A., et al., *Evaluation of matrix scaffolds for tissue engineering of articular cartilage grafts.* J Biomed Mater Res, 1997. **34**(2): p. 211-20. DOI: 10.1002/(SICI)1097-4636(199702)34:2<211::AID-JBM10>3.0.CO;2-L

[356] Middleton, J.C. and A.J. Tipton, *Synthetic biodegradable polymers as orthopedic devices.* Biomaterials, 2000. **21**(23): p. 2335-46. DOI: 10.1016/S0142-9612(00)00101-0

[357] Hutmacher, D.W., J.C. Goh, and S.H. Teoh, *An introduction to biodegradable materials for tissue engineering applications.* Ann Acad Med Singapore, 2001. **30**(2): p. 183-91.

[358] Ishaug-Riley, S.L., et al., *Human articular chondrocyte adhesion and proliferation on synthetic biodegradable polymer films.* Biomaterials, 1999. **20**(23-24): p. 2245-56. DOI: 10.1016/S0142-9612(99)00155-6

[359] Eid, K., et al., *Effect of RGD coating on osteocompatibility of PLGA-polymer disks in a rat tibial wound.* J Biomed Mater Res, 2001. **57**(2): p. 224-31. DOI: 10.1002/1097-4636(200111)57:2<224::AID-JBM1162>3.0.CO;2-F

[360] Zeltinger, J., et al., *Effect of Pore Size and Void Fraction on Cellular Adhesion, Proliferation, and Matrix Deposition.* Tissue Eng, 2001. **7**(5): p. 557-572. DOI: 10.1089/107632701753213183

[361] Honda, M., et al., *Cartilage formation by cultured chondrocytes in a new scaffold made of poly(L-lactide-epsilon-caprolactone) sponge.* J Oral Maxillofac Surg, 2000. **58**(7): p. 767-75. DOI: 10.1053/joms.2000.7262

[362] Suggs, L.J., et al., *In vitro cytotoxicity and in vivo biocompatibility of poly(propylene fumarate-co-ethylene glycol) hydrogels.* J Biomed Mater Res, 1999. **46**(1): p. 22-32. DOI: 10.1002/(SICI)1097-4636(199907)46:1<22::AID-JBM3>3.0.CO;2-R

[363] Zimmermann, J., et al., *Novel hydrogels as supports for in vitro cell growth: poly(ethylene glycol)- and gelatine-based (meth)acrylamidopeptide macromonomers.* Biomaterials, 2002. **23**(10): p. 2127-34. DOI: 10.1016/S0142-9612(01)00343-X

[364] Anseth, K.S., et al., *In situ forming degradable networks and their application in tissue engineering and drug delivery.* J Control Release, 2002. **78**(1-3): p. 199-209. DOI: 10.1016/S0168-3659(01)00500-4

[365] Fisher, J.P., et al., *Soft and hard tissue response to photocrosslinked poly(propylene fumarate) scaffolds in a rabbit model.* J Biomed Mater Res, 2002. **59**(3): p. 547-56. DOI: 10.1002/jbm.1268

[366] Suggs, L.J., et al., *Preparation and characterization of poly(propylene fumarate-co-ethylene glycol) hydrogels.* J Biomater Sci Polym Ed, 1998. **9**(7): p. 653-66. DOI: 10.1163/156856298X00073

[367] Suggs, L.J., et al., *In vitro and in vivo degradation of poly(propylene fumarate-co-ethylene glycol) hydrogels.* J Biomed Mater Res, 1998. **42**(2): p. 312-20. DOI: 10.1002/(SICI)1097-4636(199811)42:2<312::AID-JBM17>3.0.CO;2-K

[368] Suggs, L.J. and A.G. Mikos, *Development of poly(propylene fumarate-co-ethylene glycol) as an injectable carrier for endothelial cells.* Cell Transplant, 1999. **8**(4): p. 345-50.

[369] Temenoff, J.S., et al., *Effect of poly(ethylene glycol) molecular weight on tensile and swelling properties of oligo(poly(ethylene glycol) fumarate) hydrogels for cartilage tissue engineering.* J Biomed Mater Res, 2002. **59**(3): p. 429-37. DOI: 10.1002/jbm.1259

[370] Han, D.K., et al., *Surface characteristics and biocompatibility of lactide-based poly(ethylene glycol) scaffolds for tissue engineering.* J Biomater Sci Polym Ed, 1998. **9**(7): p. 667-80.

[371] Betre, H., et al., *Characterization of a genetically engineered elastin-like polypeptide for cartilaginous tissue repair.* Biomacromolecules, 2002. **3**(5): p. 910-6. DOI: 10.1021/bm0255037

[372] Ameer, G.A., T.A. Mahmood, and R. Langer, *A biodegradable composite scaffold for cell transplantation.* J Orthop Res, 2002. **20**(1): p. 16-9. DOI: 10.1016/S0736-0266(01)00074-2

[373] Marijnissen, W.J., et al., *Tissue-engineered cartilage using serially passaged articular chondrocytes. Chondrocytes in alginate, combined in vivo with a synthetic (E210) or biologic biodegradable carrier (DBM).* Biomaterials, 2000. **21**(6): p. 571-80. DOI: 10.1016/S0142-9612(99)00218-5

[374] Slivka, M.A., et al., *Porous, resorbable, fiber-reinforced scaffolds tailored for articular cartilage repair.* Tissue Eng, 2001. **7**(6): p. 767-80. DOI: 10.1089/107632701753337717

[375] Kus, W.M., et al., *Carbon fiber scaffolds in the surgical treatment of cartilage lesions.* Ann Transplant, 1999. **4**(3-4): p. 101-2.

[376] Brittberg, M., E. Faxen, and L. Peterson, *Carbon fiber scaffolds in the treatment of early knee osteoarthritis. A prospective 4-year followup of 37 patients.* Clin Orthop, 1994(307): p. 155-64.

[377] Moutos, F.T. and F. Guilak, *Composite scaffolds for cartilage tissue engineering.* Biorheology, 2008. **45**(3-4): p. 501-12.

[378] Grogan, S.P., et al., *A static, closed and scaffold-free bioreactor system that permits chondrogenesis in vitro.* Osteoarthritis Cartilage, 2003. **11**(6): p. 403-11. DOI: 10.1016/S1063-4584(03)00053-0

[379] Adkisson, H.D., et al., *In vitro generation of scaffold independent neocartilage.* Clin Orthop Relat Res, 2001(391 Suppl): p. S280-94.

[380] Boyle, J., et al., *Characterization of proteoglycan accumulation during formation of cartilagenous tissue in vitro.* Osteoarthritis Cartilage, 1995. **3**(2): p. 117-25. DOI: 10.1016/S1063-4584(05)80044-5

[381] Masuda, K., et al., *A novel two-step method for the formation of tissue-engineered cartilage by mature bovine chondrocytes: the alginate-recovered-chondrocyte (ARC) method.* J Orthop Res, 2003. **21**(1): p. 139-48. DOI: 10.1016/S0736-0266(02)00109-2

[382] Furukawa, K.S., et al., *Rapid and large-scale formation of chondrocyte aggregates by rotational culture.* Cell Transplant, 2003. **12**(5): p. 475-9.

[383] Tacchetti, C., et al., *In vitro morphogenesis of chick embryo hypertrophic cartilage.* J Cell Biol, 1987. **105**(2): p. 999-1006. DOI: 10.1083/jcb.105.2.999

[384] Gigout, A., et al., *Chondrocyte aggregation in suspension culture is GFOGER-GPP- and beta1 integrin-dependent.* J Biol Chem, 2008. **283**(46): p. 31522-30. DOI: 10.1074/jbc.M804234200

[385] Graff, R.D., S.S. Kelley, and G.M. Lee, *Role of pericellular matrix in development of a mechanically functional neocartilage.* Biotechnol Bioeng, 2003. **82**(4): p. 457-64. DOI: 10.1002/bit.10593

[386] Revell, C.M., C.E. Reynolds, and K.A. Athanasiou, *Effects of initial cell seeding in self assembly of articular cartilage.* Ann Biomed Eng, 2008. **36**(9): p. 1441-8. DOI: 10.1007/s10439-008-9524-x

[387] Elder, B.D. and K.A. Athanasiou, *Systematic assessment of growth factor treatment on biochemical and biomechanical properties of engineered articular cartilage constructs.* Osteoarthritis Cartilage, 2008. **18**: p. 18.

[388] Elder, B.D. and K.A. Athanasiou, *Effects of Temporal Hydrostatic Pressure on Tissue-Engineered Bovine Articular Cartilage Constructs.* Tissue Eng Part A, 2008. **15**(5): p. 1151-1158.

[389] Stoddart, M.J., L. Ettinger, and H.J. Hauselmann, *Enhanced matrix synthesis in de novo, scaffold free cartilage-like tissue subjected to compression and shear.* Biotechnol Bioeng, 2006. **95**(6): p. 1043-51. DOI: 10.1002/bit.21052

[390] Elder, B.D. and K.A. Athanasiou, *Synergistic and additive effects of hydrostatic pressure and growth factors on tissue formation.* PLoS One, 2008. **3**(6): p. e2341. DOI: 10.1371/journal.pone.0002341

[391] Brehm, W., et al., *Repair of superficial osteochondral defects with an autologous scaffold-free cartilage construct in a caprine model: implantation method and short-term results.* Osteoarthritis Cartilage, 2006. **14**(12): p. 1214-26. DOI: 10.1016/j.joca.2006.05.002

[392] Chaipinyo, K., B.W. Oakes, and M.P. van Damme, *Effects of growth factors on cell proliferation and matrix synthesis of low-density, primary bovine chondrocytes cultured in collagen I gels.* J Orthop Res, 2002. **20**(5): p. 1070-8. DOI: 10.1016/S0736-0266(02)00025-6

[393] Gooch, K.J., et al., *IGF-I and mechanical environment interact to modulate engineered cartilage development.* Biochem Biophys Res Commun, 2001. **286**(5): p. 909-15. DOI: 10.1006/bbrc.2001.5486

[394] Blunk, T., et al., *Differential effects of growth factors on tissue-engineered cartilage.* Tissue Eng, 2002. **8**(1): p. 73-84. DOI: 10.1089/107632702753503072

[395] Forslund, C. and P. Aspenberg, *CDMP-2 induces bone or tendon-like tissue depending on mechanical stimulation.* J Orthop Res, 2002. **20**(6): p. 1170-4. DOI: 10.1016/S0736-0266(02)00078-5

[396] Darling, E.M. and K.A. Athanasiou, *Bioactive scaffold design for articular cartilage engineering*, in *Biomedical Technology and Devices Handbook*, J. Moore and G. Zouridakis, Editors. 2004, CRC Press LLC: Boca Raton, FL. p. Chapter 21.

[397] Reddi, A.H., *Role of morphogenetic proteins in skeletal tissue engineering and regeneration.* Nat Biotechnol, 1998. **16**(3): p. 247-52. DOI: 10.1038/nbt0398-247

[398] Frenkel, S.R., et al., *Transforming growth factor beta superfamily members: role in cartilage modeling.* Plast Reconstr Surg, 2000. **105**(3): p. 980-90.

[399] Sporn, M.B., et al., *Transforming growth factor-beta: biological function and chemical structure.* Science, 1986. **233**(4763): p. 532-4.

[400] Guerne, P.A., A. Sublet, and M. Lotz, *Growth factor responsiveness of human articular chondrocytes: distinct profiles in primary chondrocytes, subcultured chondrocytes, and fibroblasts.* J Cell Physiol, 1994. **158**(3): p. 476-84. DOI: 10.1002/jcp.1041580312

[401] van der Kraan, P., E. Vitters, and W. van den Berg, *Differential effect of transforming growth factor beta on freshly isolated and cultured articular chondrocytes.* J Rheumatol, 1992. **19**(1): p. 140-5.

[402] Verschure, P.J., et al., *Responsiveness of articular cartilage from normal and inflamed mouse knee joints to various growth factors.* Ann Rheum Dis, 1994. **53**(7): p. 455-60. DOI: 10.1136/ard.53.7.455

[403] Pecina, M., et al., *Articular cartilage repair: the role of bone morphogenetic proteins.* Int Orthop, 2002. **26**(3): p. 131-6.

[404] O'Connor, W.J., et al., *The use of growth factors in cartilage repair.* Orthop Clin North Am, 2000. **31**(3): p. 399-410. DOI: 10.1016/S0030-5898(05)70159-0

[405] Gooch, K.J., et al., *Bone morphogenetic proteins-2, -12, and -13 modulate in vitro development of engineered cartilage.* Tissue Eng, 2002. **8**(4): p. 591-601. DOI: 10.1089/107632702760240517

[406] Valcourt, U., et al., *Functions of transforming growth factor-beta family type I receptors and Smad proteins in the hypertrophic maturation and osteoblastic differentiation of chondrocytes.* J Biol Chem, 2002. **277**(37): p. 33545-58. DOI: 10.1074/jbc.M202086200

[407] Luyten, F.P., et al., *Natural bovine osteogenin and recombinant human bone morphogenetic protein-2B are equipotent in the maintenance of proteoglycans in bovine articular cartilage explant cultures.* J Biol Chem, 1992. **267**(6): p. 3691-5.

[408] Mattioli-Belmonte, M., et al., *N,N-dicarboxymethyl chitosan as delivery agent for bone morphogenetic protein in the repair of articular cartilage.* Med Biol Eng Comput, 1999. **37**(1): p. 130-4. DOI: 10.1007/BF02513279

[409] Kaps, C., et al., *Bone morphogenetic proteins promote cartilage differentiation and protect engineered artificial cartilage from fibroblast invasion and destruction.* Arthritis Rheum, 2002. **46**(1): p. 149-62. DOI: 10.1002/1529-0131(200201)46:1<149::AID-ART10058>3.0.CO;2-W

[410] Erlacher, L., et al., *Presence of cartilage-derived morphogenetic proteins in articular cartilage and enhancement of matrix replacement in vitro.* Arthritis Rheum, 1998. **41**(2): p. 263-73. DOI: 10.1002/1529-0131(199802)41:2<263::AID-ART10>3.0.CO;2-5

[411] Gruber, R., et al., *Effects of cartilage-derived morphogenetic proteins and osteogenic protein-1 on osteochondrogenic differentiation of periosteum-derived cells.* Endocrinology, 2001. **142**(5): p. 2087-94. DOI: 10.1210/en.142.5.2087

[412] Pacifici, M., et al., *Development of articular cartilage: what do we know about it and how may it occur?* Connect Tissue Res, 2000. **41**(3): p. 175-84. DOI: 10.3109/03008200009005288

[413] Ziegler, J., et al., *Adsorption and release properties of growth factors from biodegradable implants.* J Biomed Mater Res, 2002. **59**(3): p. 422-8. DOI: 10.1002/jbm.1258

[414] Betre, H., et al., *A thermally responsive biopolymer for intra-articular drug delivery.* J Control Release, 2006. **115**(2): p. 175-82. DOI: 10.1016/j.jconrel.2006.07.022

[415] Madry, H., et al., *Gene transfer of a human insulin-like growth factor I cDNA enhances tissue engineering of cartilage.* Hum Gene Ther, 2002. **13**(13): p. 1621-30. DOI: 10.1089/10430340260201716

[416] Grimaud, E., D. Heymann, and F. Redini, *Recent advances in TGF-beta effects on chondrocyte metabolism. Potential therapeutic roles of TGF-beta in cartilage disorders.* Cytokine Growth Factor Rev, 2002. **13**(3): p. 241-57. DOI: 10.1016/S1359-6101(02)00004-7

[417] Elisseeff, J., et al., *Photoencapsulation of chondrocytes in poly(ethylene oxide)-based semi- interpenetrating networks.* J Biomed Mater Res, 2000. **51**(2): p. 164-71. DOI: 10.1002/(SICI)1097-4636(200008)51:2<164::AID-JBM4>3.0.CO;2-W

[418] Mann, B.K., R.H. Schmedlen, and J.L. West, *Tethered-TGF-beta increases extracellular matrix production of vascular smooth muscle cells.* Biomaterials, 2001. **22**(5): p. 439-44. DOI: 10.1016/S0142-9612(00)00196-4

[419] Fan, V.H., et al., *Tethered epidermal growth factor provides a survival advantage to mesenchymal stem cells.* Stem Cells, 2007. **25**(5): p. 1241-51. DOI: 10.1634/stemcells.2006-0320

[420] LeBaron, R.G. and K.A. Athanasiou, *Extracellular matrix cell adhesion peptides: functional applications in orthopedic materials.* Tissue Eng, 2000. **6**(2): p. 85-103. DOI: 10.1089/107632700320720

[421] Woods, V.L., Jr., et al., *Integrin expression by human articular chondrocytes.* Arthritis Rheum, 1994. **37**(4): p. 537-44. DOI: 10.1002/art.1780370414

[422] Shimizu, M., et al., *Chondrocyte migration to fibronectin, type I collagen, and type II collagen.* Cell Struct Funct, 1997. **22**(3): p. 309-15.

[423] Bhati, R.S., et al., *The growth of chondrocytes into a fibronectin-coated biodegradable scaffold.* J Biomed Mater Res, 2001. **56**(1): p. 74-82. DOI: 10.1002/1097-4636(200107)56:1<74::AID-JBM1070>3.0.CO;2-M

[424] Solchaga, L.A., et al., *Treatment of osteochondral defects with autologous bone marrow in a hyaluronan-based delivery vehicle.* Tissue Eng, 2002. **8**(2): p. 333-47. DOI: 10.1089/107632702753725085

[425] Prime, K.L. and G.M. Whitesides, *Self-assembled organic monolayers: model systems for studying adsorption of proteins at surfaces.* Science, 1991. **252**(5010): p. 1164-7.

[426] Ruoslahti, E., *RGD and other recognition sequences for integrins.* Annu Rev Cell Dev Biol, 1996. **12**: p. 697-715. DOI: 10.1146/annurev.cellbio.12.1.697

[427] Hubbell, J.A., *Matrix Effects*, in *Principles of Tissue Engineering*, R. Lanza, R. Langer, and W. Chick, Editors. 1997, R. G. Landes Company: Austin. p. 249.

[428] Lee, V., et al., *The roles of matrix molecules in mediating chondrocyte aggregation, attachment, and spreading.* J Cell Biochem, 2000. **79**(2): p. 322-33. DOI: 10.1002/1097-4644(20001101)79:2<322::AID-JCB150>3.0.CO;2-U

[429] Jennings, L., et al., *The effects of collagen fragments on the extracellular matrix metabolism of bovine and human chondrocytes.* Connect Tissue Res, 2001. **42**(1): p. 71-86. DOI: 10.3109/03008200109014250

[430] Thomas, C.H., et al., *The role of vitronectin in the attachment and spatial distribution of bone-derived cells on materials with patterned surface chemistry.* J Biomed Mater Res, 1997. **37**(1): p. 81-93. DOI: 10.1002/(SICI)1097-4636(199710)37:1<81::AID-JBM10>3.0.CO;2-T

[431] Wyre, R.M. and S. Downes, *The role of protein adsorption on chondrocyte adhesion to a heterocyclic methacrylate polymer system.* Biomaterials, 2002. **23**(2): p. 357-64. DOI: 10.1016/S0142-9612(01)00113-2

[432] Choi, H.U., et al., *Isolation and characterization of a 35,000 molecular weight subunit fetal cartilage matrix protein.* J Biol Chem, 1983. **258**(1): p. 655-61.

[433] Makihira, S., et al., *Enhancement of cell adhesion and spreading by a cartilage-specific noncollagenous protein, cartilage matrix protein (CMP/Matrilin-1), via integrin alpha1beta1.* J Biol Chem, 1999. **274**(16): p. 11417-23. DOI: 10.1074/jbc.274.16.11417

[434] Gallant, N.D., et al., *Micropatterned Surfaces to Engineer Focal Adhesions for Analysis of Cell Adhesion Strengthening.* LANGMUIR, 2002. **18**: p. 5579-5584. DOI: 10.1021/la025554p

[435] Koivunen, E., et al., *Peptides in cell adhesion research.* Methods Enzymol, 1994. **245**: p. 346-69. DOI: 10.1016/0076-6879(94)45019-6

[436] Pasqualini, R., E. Koivunen, and E. Ruoslahti, *Peptides in cell adhesion: powerful tools for the study of integrin- ligand interactions.* Braz J Med Biol Res, 1996. **29**(9): p. 1151-8.

[437] Jo, S., P.S. Engel, and A.G. Mikos, *Synthesis of poly(ethylene glycol)-tethered poly(propylene fumarate) and its modification with GRGD peptide.* Polymer, 2000. **41**: p. 7595-7604. DOI: 10.1016/S0032-3861(00)00117-8

[438] Pierschbacher, M.D. and E. Ruoslahti, *Variants of the cell recognition site of fibronectin that retain attachment-promoting activity.* Proc Natl Acad Sci U S A, 1984. **81**(19): p. 5985-8. DOI: 10.1073/pnas.81.19.5985

[439] Rezania, A., et al., *The detachment strength and morphology of bone cells contacting materials modified with a peptide sequence found within bone sialoprotein.* J Biomed Mater Res, 1997. **37**(1): p. 9-19. DOI: 10.1002/(SICI)1097-4636(199710)37:1<9::AID-JBM2>3.0.CO;2-W

[440] Petersen, E.F., R.G. Spencer, and E.W. McFarland, *Microengineering neocartilage scaffolds.* Biotechnol Bioeng, 2002. **78**(7): p. 801-4. DOI: 10.1002/bit.10256

[441] Mann, B.K., et al., *Smooth muscle cell growth in photopolymerized hydrogels with cell adhesive and proteolytically degradable domains: synthetic ECM analogs for tissue engineering.* Biomaterials, 2001. **22**(22): p. 3045-51. DOI: 10.1016/S0142-9612(01)00051-5

[442] Gobin, A.S. and J.L. West, *Cell migration through defined, synthetic ECM analogs.* Faseb J, 2002. **16**(7): p. 751-3.

[443] Alsberg, E., et al., *Cell-interactive alginate hydrogels for bone tissue engineering.* J Dent Res, 2001. **80**(11): p. 2025-9.

[444] Mann, B.K. and J.L. West, *Cell adhesion peptides alter smooth muscle cell adhesion, proliferation, migration, and matrix protein synthesis on modified surfaces and in polymer scaffolds.* J Biomed Mater Res, 2002. **60**(1): p. 86-93. DOI: 10.1002/jbm.10042

[445] Mann, B.K., et al., *Modification of surfaces with cell adhesion peptides alters extracellular matrix deposition.* Biomaterials, 1999. **20**(23-24): p. 2281-6. DOI: 10.1016/S0142-9612(99)00158-1

[446] Britland, S., et al., *Micropatterned substratum adhesiveness: a model for morphogenetic cues controlling cell behavior.* Exp Cell Res, 1992. **198**(1): p. 124-9. DOI: 10.1016/0014-4827(92)90157-4

[447] Lom, B., K.E. Healy, and P.E. Hockberger, *A versatile technique for patterning biomolecules onto glass coverslips.* J Neurosci Methods, 1993. **50**(3): p. 385-97. DOI: 10.1016/0165-0270(93)90044-R

[448] Dike, L.E., et al., *Geometric control of switching between growth, apoptosis, and differentiation during angiogenesis using micropatterned substrates.* In Vitro Cell Dev Biol Anim, 1999. **35**(8): p. 441-8. DOI: 10.1007/s11626-999-0050-4

[449] Chen, C.S., et al., *Geometric control of cell life and death.* Science, 1997. **276**(5317): p. 1425-8.

[450] Lee, J.Y., et al., *Analysis of local tissue-specific gene expression in cellular micropatterns.* Anal Chem, 2006. **78**(24): p. 8305-12. DOI: 10.1021/ac0613333

[451] Bettinger, C.J., R. Langer, and J.T. Borenstein, *Engineering Substrate Topography at the Micro- and Nanoscale to Control Cell Function.* Angew Chem Int Ed Engl, 2009.

[452] Yim, E.K., S.W. Pang, and K.W. Leong, *Synthetic nanostructures inducing differentiation of human mesenchymal stem cells into neuronal lineage.* Exp Cell Res, 2007. **313**(9): p. 1820-9. DOI: 10.1016/j.yexcr.2007.02.031

[453] Natoli, R., C.M. Revell, and K. Athanasiou, *Chondroitinase ABC Treatment Results in Increased Tensile Properties of Self-Assembled Tissue Engineered Articular Cartilage.* Tissue Eng Part A, 2009: p. doi:10.1089/ten.TEA.2008.0478. DOI: 10.1089/ten.tea.2008.0478

[454] Natoli, R.M., et al., *Effects of multiple chondroitinase ABC applications on tissue engineered articular cartilage.* J Orthop Res, 2009. DOI: 10.1002/jor.20821

[455] Siegel, R.C., S.R. Pinnell, and G.R. Martin, *Cross-linking of collagen and elastin. Properties of lysyl oxidase.* Biochemistry, 1970. **9**(23): p. 4486-92. DOI: 10.1021/bi00825a004

[456] Asanbaeva, A., et al., *Cartilage growth and remodeling: modulation of balance between proteoglycan and collagen network in vitro with beta-aminopropionitrile.* Osteoarthritis Cartilage, 2008. **16**(1): p. 1-11. DOI: 10.1016/j.joca.2007.05.019

[457] McGowan, K.B. and R.L. Sah, *Treatment of cartilage with beta-aminopropionitrile accelerates subsequent collagen maturation and modulates integrative repair.* J Orthop Res, 2005. **23**(3): p. 594-601. DOI: 10.1016/j.orthres.2004.02.015

[458] Wong, M., et al., *Collagen fibrillogenesis by chondrocytes in alginate.* Tissue Eng, 2002. **8**(6): p. 979-87. DOI: 10.1089/107632702320934074

[459] Dunkelman, N.S., et al., *Cartilage production by rabbit articular chondrocytes on polyglycolic acid scaffolds in a closed bioreactor system.* Biotechnol Bioeng, 1995. **46**: p. 299-305. DOI: 10.1002/bit.260460402

[460] Vunjak-Novakovic, G., et al., *Bioreactor cultivation conditions modulate the composition and mechanical properties of tissue-engineered cartilage.* J Orthop Res, 1999. **17**(1): p. 130-8. DOI: 10.1002/jor.1100170119

[461] Schulz, R.M., et al., *Development and validation of a novel bioreactor system for load- and perfusion-controlled tissue engineering of chondrocyte-constructs.* Biotechnol Bioeng, 2008. **101**(4): p. 714-28. DOI: 10.1002/bit.21955

[462] Demarteau, O., et al., *Dynamic compression of cartilage constructs engineered from expanded human articular chondrocytes.* Biochem Biophys Res Commun, 2003. **310**(2): p. 580-8. DOI: 10.1016/j.bbrc.2003.09.099

[463] Palmoski, M.J. and K.D. Brandt, *Effects of static and cyclic compressive loading on articular cartilage plugs in vitro.* Arthritis Rheum, 1984. **27**(6): p. 675-81. DOI: 10.1002/art.1780270611

[464] Kim, Y.J., et al., *Mechanical regulation of cartilage biosynthetic behavior: physical stimuli.* Arch Biochem Biophys, 1994. **311**(1): p. 1-12. DOI: 10.1006/abbi.1994.1201

[465] Buschmann, M.D., et al., *Mechanical compression modulates matrix biosynthesis in chondrocyte/agarose culture.* J Cell Sci, 1995. **108**(Pt 4): p. 1497-508.

[466] Torzilli, P.A., et al., *Characterization of cartilage metabolic response to static and dynamic stress using a mechanical explant test system.* J Biomech, 1997. **30**(1): p. 1-9. DOI: 10.1016/S0021-9290(96)00117-0

[467] Davisson, T., et al., *Static and dynamic compression modulate matrix metabolism in tissue engineered cartilage.* J Orthop Res, 2002. **20**(4): p. 842-8. DOI: 10.1016/S0736-0266(01)00160-7

[468] Wang, Q.G., et al., *Gene expression profiles of dynamically compressed single chondrocytes and chondrons.* Biochem Biophys Res Commun, 2009. **379**(3): p. 738-42. DOI: 10.1016/j.bbrc.2008.12.111

[469] Lee, D.A., et al., *The influence of mechanical loading on isolated chondrocytes seeded in agarose constructs.* Biorheology, 2000. **37**(1-2): p. 149-61.

[470] Sah, R.L., et al., *Biosynthetic response of cartilage explants to dynamic compression.* J Orthop Res, 1989. **7**(5): p. 619-36. DOI: 10.1002/jor.1100070502

[471] Campbell, J.J., D.A. Lee, and D.L. Bader, *Dynamic compressive strain influences chondrogenic gene expression in human mesenchymal stem cells.* Biorheology, 2006. **43**(3-4): p. 455-70.

[472] Mauck, R.L., X. Yuan, and R.S. Tuan, *Chondrogenic differentiation and functional maturation of bovine mesenchymal stem cells in long-term agarose culture.* Osteoarthritis Cartilage, 2006. **14**(2): p. 179-89. DOI: 10.1016/j.joca.2005.09.002

[473] Bougault, C., et al., *Molecular analysis of chondrocytes cultured in agarose in response to dynamic compression.* BMC Biotechnol, 2008. **8**: p. 71. DOI: 10.1186/1472-6750-8-71

[474] Wang, P.Y., et al., *Modulation of gene expression of rabbit chondrocytes by dynamic compression in polyurethane scaffolds with collagen gel encapsulation.* J Biomater Appl, 2009. **23**(4): p. 347-66.

[475] Pelaez, D., C.Y. Huang, and H.S. Cheung, *Cyclic Compression Maintains Viability and Induces Chondrogenesis of Human Mesenchymal Stem Cells in Fibrin Gel Scaffolds.* Stem Cells Dev, 2008.

[476] Jung, Y., et al., *Cartilaginous tissue formation using a mechano-active scaffold and dynamic compressive stimulation.* J Biomater Sci Polym Ed, 2008. **19**(1): p. 61-74. DOI: 10.1163/156856208783227712

[477] Kisiday, J.D., et al., *Effects of dynamic compressive loading on chondrocyte biosynthesis in self-assembling peptide scaffolds.* J Biomech, 2004. **37**(5): p. 595-604. DOI: 10.1016/j.jbiomech.2003.10.005

[478] Lima, E.G., et al., *The beneficial effect of delayed compressive loading on tissue-engineered cartilage constructs cultured with TGF-beta3.* Osteoarthritis Cartilage, 2007. **15**(9): p. 1025-33. DOI: 10.1016/j.joca.2007.03.008

[479] Kelly, T.A., et al., *Analysis of radial variations in material properties and matrix composition of chondrocyte-seeded agarose hydrogel constructs.* Osteoarthritis Cartilage, 2009. **17**(1): p. 73-82. DOI: 10.1016/j.joca.2008.05.019

[480] Chowdhury, T.T., et al., *Temporal regulation of chondrocyte metabolism in agarose constructs subjected to dynamic compression.* Arch Biochem Biophys, 2003. **417**(1): p. 105-11. DOI: 10.1016/S0003-9861(03)00340-0

[481] Liu, T.Y., et al., *[Influence of transforming growth factor-beta1 inducing time on chondrogenesis of bone marrow stromal cells (BMSCs): in vitro experiment with porcine BMSCs].* Zhonghua Yi Xue Za Zhi, 2007. **87**(31): p. 2218-22.

[482] Chowdhury, T.T., et al., *Integrin-mediated mechanotransduction processes in TGFbeta-stimulated monolayer-expanded chondrocytes.* Biochem Biophys Res Commun, 2004. **318**(4): p. 873-81. DOI: 10.1016/j.bbrc.2004.04.107

[483] Suh, J.K., *Dynamic unconfined compression of articular cartilage under a cyclic compressive load.* Biorheology, 1996. **33**(4-5): p. 289-304. DOI: 10.1016/0006-355X(96)00023-6

[484] Mauck, R.L., C.T. Hung, and G.A. Ateshian, *Modeling of neutral solute transport in a dynamically loaded porous permeable gel: implications for articular cartilage biosynthesis and tissue engineering.* J Biomech Eng, 2003. **125**(5): p. 602-14. DOI: 10.1115/1.1611512

[485] Hall, A.C., E.R. Horwitz, and R.J. Wilkins, *The cellular physiology of articular cartilage.* Exp Physiol, 1996. **81**(3): p. 535-45.

[486] Bachrach, N.M., V.C. Mow, and F. Guilak, *Incompressibility of the solid matrix of articular cartilage under high hydrostatic pressures.* J Biomech, 1998. **31**(5): p. 445-51. DOI: 10.1016/S0021-9290(98)00035-9

[487] van Kampen, G.P., et al., *Cartilage response to mechanical force in high-density chondrocyte cultures.* Arthritis Rheum, 1985. **28**(4): p. 419-24.

[488] Veldhuijzen, J.P., et al., *The growth of cartilage cells in vitro and the effect of intermittent compressive force. A histological evaluation.* Connect Tissue Res, 1987. **16**(2): p. 187-96. DOI: 10.3109/03008208709002006

[489] Smith, R.L., et al., *In vitro stimulation of articular chondrocyte mRNA and extracellular matrix synthesis by hydrostatic pressure.* J Orthop Res, 1996. **14**(1): p. 53-60. DOI: 10.1002/jor.1100140110

[490] Smith, R.L., et al., *Time-dependent effects of intermittent hydrostatic pressure on articular chondrocyte type II collagen and aggrecan mRNA expression.* J Rehabil Res Dev, 2000. **37**(2): p. 153-61.

[491] Hansen, U., et al., *Combination of reduced oxygen tension and intermittent hydrostatic pressure: a useful tool in articular cartilage tissue engineering.* J Biomech, 2001. **34**: p. 941-49. DOI: 10.1016/S0021-9290(01)00050-1

[492] Parkkinen, J.J., et al., *Effects of cyclic hydrostatic pressure on proteoglycan synthesis in cultured chondrocytes and articular cartilage explants.* Arch Biochem Biophys, 1993. **300**(1): p. 458-65. DOI: 10.1006/abbi.1993.1062

[493] Carver, S.E. and C.A. Heath, *Semi-continuous perfusion system for delivering intermittent physiological pressure to regenerating cartilage.* Tissue Eng, 1999. **5**(1): p. 1-11. DOI: 10.1089/ten.1999.5.1

[494] Heath, C.A. and S.R. Magari, *Mini-review: Mechanical factors affecting cartilage regeneration in vitro.* Biotechnol Bioeng, 1996. **50**: p. 430-437. DOI: 10.1002/(SICI)1097-0290(19960520)50:4<430::AID-BIT10>3.0.CO;2-N

[495] Carver, S.E. and C.A. Heath, *Increasing extracellular matrix production in regenerating cartilage with intermittent physiological pressure.* Biotechnol Bioeng, 1999. **62**(2): p. 166-74. DOI: 10.1002/(SICI)1097-0290(19990120)62:2<166::AID-BIT6>3.0.CO;2-K

[496] Lammi, M.J., et al., *Expression of reduced amounts of structurally altered aggrecan in articular cartilage chondrocytes exposed to high hydrostatic pressure.* Biochem J, 1994. **304**(Pt 3): p. 723-30.

[497] Takahashi, K., et al., *Hydrostatic pressure induces expression of interleukin 6 and tumour necrosis factor alpha mRNAs in a chondrocyte-like cell line.* Ann Rheum Dis, 1998. **57**(4): p. 231-6. DOI: 10.1136/ard.57.4.231

[498] Domm, C., et al., *[Redifferentiation of dedifferentiated joint cartilage cells in alginate culture. Effect of intermittent hydrostatic pressure and low oxygen partial pressure].* Orthopade, 2000. **29**(2): p. 91-9.

[499] Hall, A.C., J.P. Urban, and K.A. Gehl, *The effects of hydrostatic pressure on matrix synthesis in articular cartilage.* J Orthop Res, 1991. **9**(1): p. 1-10. DOI: 10.1002/jor.1100090102

[500] Sironen, R., et al., *Transcriptional activation in chondrocytes submitted to hydrostatic pressure.* Biorheology, 2000. **37**(1-2): p. 85-93.

[501] Kaarniranta, K., et al., *Hsp70 accumulation in chondrocytic cells exposed to high continuous hydrostatic pressure coincides with mRNA stabilization rather than transcriptional activation.* Proc Natl Acad Sci U S A, 1998. **95**(5): p. 2319-24. DOI: 10.1073/pnas.95.5.2319

[502] Parkkinen, J.J., et al., *Influence of short-term hydrostatic pressure on organization of stress fibers in cultured chondrocytes.* J Orthop Res, 1995. **13**(4): p. 495-502. DOI: 10.1002/jor.1100130404

[503] Parkkinen, J.J., et al., *Altered Golgi apparatus in hydrostatically loaded articular cartilage chondrocytes.* Ann Rheum Dis, 1993. **52**(3): p. 192-8. DOI: 10.1136/ard.52.3.192

[504] Nakamura, S., et al., *Hydrostatic pressure induces apoptosis of chondrocytes cultured in alginate beads.* J Orthop Res, 2006. **24**(4): p. 733-9. DOI: 10.1002/jor.20077

[505] Wenger, R., et al., *Hydrostatic pressure increases apoptosis in cartilage-constructs produced from human osteoarthritic chondrocytes.* Front Biosci, 2006. **11**: p. 1690-5. DOI: 10.2741/1914

[506] Ogawa, R., et al., *The Effect of Hydrostatic Pressure on 3-D Chondroinduction of Human Adipose-Derived Stem Cells.* Tissue Eng Part A, 2009. DOI: 10.1089/ten.tea.2008.0672

[507] Clark, C.C., B.S. Tolin, and C.T. Brighton, *The effect of oxygen tension on proteoglycan synthesis and aggregation in mammalian growth plate chondrocytes.* Journal of Orthopaedic Research, 1991. **9**(4): p. 477-84. DOI: 10.1002/jor.1100090403

[508] Candiani, G., et al., *Chondrocyte response to high regimens of cyclic hydrostatic pressure in 3-dimensional engineered constructs.* Int J Artif Organs, 2008. **31**(6): p. 490-9.

[509] Heyland, J., et al., *Redifferentiation of chondrocytes and cartilage formation under intermittent hydrostatic pressure.* Biotechnol Lett, 2006. **28**(20): p. 1641-8. DOI: 10.1007/s10529-006-9144-1

[510] Kawanishi, M., et al., *Redifferentiation of dedifferentiated bovine articular chondrocytes enhanced by cyclic hydrostatic pressure under a gas-controlled system.* Tissue Eng, 2007. **13**(5): p. 957-64. DOI: 10.1089/ten.2006.0176

[511] Luo, Z.J. and B.B. Seedhom, *Light and low-frequency pulsatile hydrostatic pressure enhances extracellular matrix formation by bone marrow mesenchymal cells: an in-vitro study with special reference to cartilage repair.* Proc Inst Mech Eng [H], 2007. **221**(5): p. 499-507.

[512] Miyanishi, K., et al., *Dose- and time-dependent effects of cyclic hydrostatic pressure on transforming growth factor-beta3-induced chondrogenesis by adult human mesenchymal stem cells in vitro.* Tissue Eng, 2006. **12**(8): p. 2253-62. DOI: 10.1089/ten.2006.12.2253

[513] Miyanishi, K., et al., *Effects of hydrostatic pressure and transforming growth factor-beta 3 on adult human mesenchymal stem cell chondrogenesis in vitro.* Tissue Eng, 2006. **12**(6): p. 1419-28. DOI: 10.1089/ten.2006.12.1419

[514] Sakao, K., et al., *Induction of chondrogenic phenotype in synovium-derived progenitor cells by intermittent hydrostatic pressure.* Osteoarthritis Cartilage, 2008. **16**(7): p. 805-14. DOI: 10.1016/j.joca.2007.10.021

[515] Wagner, D.R., et al., *Hydrostatic pressure enhances chondrogenic differentiation of human bone marrow stromal cells in osteochondrogenic medium.* Ann Biomed Eng, 2008. **36**(5): p. 813-20. DOI: 10.1007/s10439-008-9448-5

[516] Elder, S.H., et al., *Influence of hydrostatic and distortional stress on chondroinduction.* Biorheology, 2008. **45**(3-4): p. 479-86.

[517] Sharma, G., R.K. Saxena, and P. Mishra, *Synergistic effect of chondroitin sulfate and cyclic pressure on biochemical and morphological properties of chondrocytes from articular cartilage.* Osteoarthritis Cartilage, 2008. **16**(11): p. 1387-94. DOI: 10.1016/j.joca.2008.03.026

[518] Wimmer, M.A., et al., *Tribology approach to the engineering and study of articular cartilage.* Tissue Eng, 2004. **10**(9-10): p. 1436-45. DOI: 10.1089/ten.2004.10.1436

[519] Waldman, S.D., et al., *Long-term intermittent shear deformation improves the quality of cartilaginous tissue formed in vitro.* J Orthop Res, 2003. **21**(4): p. 590-6. DOI: 10.1016/S0736-0266(03)00009-3

[520] Frank, E.H., et al., *A versatile shear and compression apparatus for mechanical stimulation of tissue culture explants.* J Biomech, 2000. **33**(11): p. 1523-7. DOI: 10.1016/S0021-9290(00)00100-7

[521] Jin, M., et al., *Combined effects of dynamic tissue shear deformation and insulin-like growth factor I on chondrocyte biosynthesis in cartilage explants.* Arch Biochem Biophys, 2003. **414**(2): p. 223-31. DOI: 10.1016/S0003-9861(03)00195-4

[522] Jin, M., et al., *Tissue shear deformation stimulates proteoglycan and protein biosynthesis in bovine cartilage explants.* Arch Biochem Biophys, 2001. **395**(1): p. 41-8. DOI: 10.1006/abbi.2001.2543

[523] Gooch, K.J., et al., *Effects of mixing intensity on tissue-engineered cartilage.* Biotechnol Bioeng, 2001. **72**(4): p. 402-7. DOI: 10.1002/1097-0290(20000220)72:4<402::AID-BIT1002>3.0.CO;2-Q

[524] Bueno, E.M., B. Bilgen, and G.A. Barabino, *Wavy-walled bioreactor supports increased cell proliferation and matrix deposition in engineered cartilage constructs.* Tissue Eng, 2005. **11**(11-12): p. 1699-709. DOI: 10.1089/ten.2005.11.1699

[525] Vunjak-Novakovic, G., et al., *Dynamic cell seeding of polymer scaffolds for cartilage tissue engineering.* Biotechnol Prog, 1998. **14**(2): p. 193-202. DOI: 10.1021/bp970120j

[526] Furukawa, K.S., et al., *Scaffold-free cartilage by rotational culture for tissue engineering.* J Biotechnol, 2008. **133**(1): p. 134-45. DOI: 10.1016/j.jbiotec.2007.07.957

[527] Vunjak-Novakovic, G., et al., *Effects of mixing on the composition and morphology of tissue-engineered cartilage.* AIChE Journal, 1996. **42**(3): p. 850-60. DOI: 10.1002/aic.690420323

[528] Cherry, R.S. and T. Papoutsakis, *Understanding and controlling fluid-mechanical injury of animal cells in bioreactors*, in *Animal Cell Biotechnology*, R.E. Spier and J.B. Griffiths, Editors. 1990, Academic Press: San Diego.

[529] Freed, L.E., et al., *Composition of cell-polymer cartilage implants.* Biotechnol Bioeng, 1994. **43**: p. 605-14. DOI: 10.1002/bit.260430710

[530] Merchuk, J.C., *Why use air-lift bioreactors?* Trends Biotechnol, 1990. **8**: p. 66-71. DOI: 10.1016/0167-7799(90)90138-N

[531] Bussolari, S., C. Dewey, and M. Gimbrone, *Apparatus for subjecting living cells to fluid shear shear stress.* Review of Scientific Instruments, 1982. **53**(12): p. 1851-54. DOI: 10.1063/1.1136909

[532] Lee, M.S., et al., *Effects of shear stress on nitric oxide and matrix protein gene expression in human osteoarthritic chondrocytes in vitro.* J Orthop Res, 2002. **20**(3): p. 556-61. DOI: 10.1016/S0736-0266(01)00149-8

[533] Smith, R.L., et al., *Effects of shear stress on articular chondrocyte metabolism.* Biorheology, 2000. **37**(1-2): p. 95-107.

[534] Mizuno, S., F. Allemann, and J. Glowacki, *Effects of medium perfusion on matrix production by bovine chondrocytes in three-dimensional collagen sponges.* J Biomed Mater Res, 2001. **56**(3): p. 368-75. DOI: 10.1002/1097-4636(20010905)56:3<368::AID-JBM1105>3.0.CO;2-V

[535] Davisson, T., R.L. Sah, and A. Ratcliffe, *Perfusion increases cell content and matrix synthesis in chondrocyte three-dimensional cultures.* Tissue Eng, 2002. **8**(5): p. 807-16. DOI: 10.1089/10763270260424169

[536] Khan, A.A., et al., *The effect of continuous culture on the growth and structure of tissue-engineered cartilage.* Biotechnol Prog, 2009. DOI: 10.1002/btpr.108

[537] Sittinger, M., et al., *Engineering of cartilage tissue using bioresorbable polymer carriers in perfusion culture.* Biomaterials, 1994. **15**(6): p. 451-6. DOI: 10.1016/0142-9612(94)90224-0

[538] Pazzano, D., et al., *Comparison of chondrogensis in static and perfused bioreactor culture.* Biotechnol Prog, 2000. **16**(5): p. 893-6. DOI: 10.1021/bp000082v

[539] Wendt, D., et al., *Uniform tissues engineered by seeding and culturing cells in 3D scaffolds under perfusion at defined oxygen tensions.* Biorheology, 2006. **43**(3-4): p. 481-8.

[540] Raimondi, M.T., et al., *Engineered cartilage constructs subject to very low regimens of interstitial perfusion.* Biorheology, 2008. **45**(3-4): p. 471-8.

[541] Raimondi, M.T., et al., *The effect of hydrodynamic shear on 3D engineered chondrocyte systems subject to direct perfusion.* Biorheology, 2006. **43**(3-4): p. 215-22.

[542] Goodwin, T.J., et al., *Reduced shear stress: a major component in the ability of mammalian tissues to form three-dimensional assemblies in simulated microgravity.* J Cell Biochem, 1993. **51**(3): p. 301-11. DOI: 10.1002/jcb.240510309

[543] Tsao, Y.D. and S.R. Gonda, *A new technology for three-dimensional cell culture: the hydrodynamic focusing bioreactor.* Advances in Heat and Mass Transfer in Biotechology, 1999. **44**: p. 37-38.

[544] Newcombe, F.C., *Limitations of the Klinostat as an Instrument for Scientific Research.* Science, 1904. **20**(507): p. 376-379. DOI: 10.1126/science.20.507.376-b

[545] Briegleb, W. *The clinostat — a tool for analyzing the influence of acceleration on solid-liquid systems.* in *Workshop on Space Biology.* 1983. Cologne, Germany.

[546] Obradovic, B., et al., *Bioreactor studies of natural and tissue engineered cartilage.* Ortop Traumatol Rehabil, 2001. **3**(2): p. 181-9.

[547] Tognana, E., et al., *Adjacent tissues (cartilage, bone) affect the functional integration of engineered calf cartilage in vitro.* Osteoarthritis Cartilage, 2005. **13**(2): p. 129-38. DOI: 10.1016/j.joca.2004.10.015

[548] Duke, P.J., E.L. Daane, and D. Montufar-Solis, *Studies of chondrogenesis in rotating systems.* J Cell Biochem, 1993. **51**(3): p. 274-82. DOI: 10.1002/jcb.240510306

[549] Freed, L.E. and G. Vunjak-Novakovic, *Cultivation of cell-polymer tissue constructs in simulated microgravity.* Biotechnol Bioeng, 1995. **46**: p. 306. DOI: 10.1002/bit.260460403

[550] Martin, I., et al., *Method for quantitative analysis of glycosaminoglycan distribution in cultured natural and engineered cartilage.* Ann Biomed Eng, 1999. **27**(5): p. 656-62. DOI: 10.1114/1.205

[551] Martin, I., et al., *Modulation of the mechanical properties of tissue engineered cartilage.* Biorheology, 2000. **37**(1-2): p. 141-7.

[552] Begley, C.M. and S.J. Kleis, *The fluid dynamic and shear environment in the NASA/JSC rotating-wall perfused-vessel bioreactor.* Biotechnol Bioeng, 2000. **70**(1): p. 32-40. DOI: 10.1002/1097-0290(20001005)70:1<32::AID-BIT5>3.0.CO;2-V

[553] Tsao, Y.-M.D., et al., *Fluid dynamics within a rotating bioreactor in space and earth environments.* Journal of Spacecraft and Rockets, 1994. **31**(6): p. 937-43. DOI: 10.2514/3.26541

[554] Tsao, Y.-M.D., S.R. Gonda, and N.R. Pellis, *Mass transfer characteristics of NASA biorectors by numerical simulation.* Advances in Heat and Mass Transfer in Biotechnology, 1997. **37**.

[555] Akmal, M., et al., *The culture of articular chondrocytes in hydrogel constructs within a bioreactor enhances cell proliferation and matrix synthesis.* J Bone Joint Surg Br, 2006. **88**(4): p. 544-53.

[556] Freed, L.E., I. Martin, and G. Vunjak-Novakovic, *Frontiers in tissue engineering. In vitro modulation of chondrogenesis.* Clin Orthop, 1999(367 Suppl): p. S46-58.

[557] Obradovic, B., et al., *Gas exchange is essential for bioreactor cultivation of tissue engineered cartilage.* Biotechnol Bioeng, 1999. **63**(2): p. 197-205. DOI: 10.1002/(SICI)1097-0290(19990420)63:2<197::AID-BIT8>3.0.CO;2-2

[558] Villanueva, I., et al., *Cross-linking density alters early metabolic activities in chondrocytes encapsulated in poly(ethylene glycol) hydrogels and cultured in the rotating wall vessel.* Biotechnol Bioeng, 2009. **102**(4): p. 1242-50. DOI: 10.1002/bit.22134

[559] Chen, H.C., et al., *Combination of baculovirus-expressed BMP-2 and rotating-shaft bioreactor culture synergistically enhances cartilage formation.* Gene Ther, 2008. **15**(4): p. 309-17. DOI: 10.1038/sj.gt.3303087

[560] Chen, H.C., et al., *The repair of osteochondral defects using baculovirus-mediated gene transfer with de-differentiated chondrocytes in bioreactor culture.* Biomaterials, 2009. **30**(4): p. 674-81. DOI: 10.1016/j.biomaterials.2008.10.017

[561] Marlovits, S., et al., *Chondrogenesis of aged human articular cartilage in a scaffold-free bioreactor.* Tissue Eng, 2003. **9**(6): p. 1215-26. DOI: 10.1089/10763270360728125

[562] Marlovits, S., et al., *Collagen expression in tissue engineered cartilage of aged human articular chondrocytes in a rotating bioreactor.* Int J Artif Organs, 2003. **26**(4): p. 319-30.

[563] Sakai, S., et al., *Rotating three-dimensional dynamic culture of adult human bone marrow-derived cells for tissue engineering of hyaline cartilage.* J Orthop Res, 2009. **27**(4): p. 517-21. DOI: 10.1002/jor.20566

[564] Augst, A., et al., *Effects of chondrogenic and osteogenic regulatory factors on composite constructs grown using human mesenchymal stem cells, silk scaffolds and bioreactors.* J R Soc Interface, 2008. **5**(25): p. 929-39. DOI: 10.1098/rsif.2007.1302

[565] Ohyabu, Y., et al., *Cartilaginous tissue formation from bone marrow cells using rotating wall vessel (RWV) bioreactor.* Biotechnol Bioeng, 2006. **95**(5): p. 1003-8. DOI: 10.1002/bit.20892

[566] Pound, J.C., et al., *Strategies to promote chondrogenesis and osteogenesis from human bone marrow cells and articular chondrocytes encapsulated in polysaccharide templates.* Tissue Eng, 2006. **12**(10): p. 2789-99. DOI: 10.1089/ten.2006.12.2789

[567] Pei, M., et al., *Engineering of functional cartilage tissue using stem cells from synovial lining: a preliminary study.* Clin Orthop Relat Res, 2008. **466**(8): p. 1880-9. DOI: 10.1007/s11999-008-0316-2

[568] Pei, M., et al., *Repair of full-thickness femoral condyle cartilage defects using allogeneic synovial cell-engineered tissue constructs.* Osteoarthritis Cartilage, 2008.

[569] Kunisaki, S.M., R.W. Jennings, and D.O. Fauza, *Fetal cartilage engineering from amniotic mesenchymal progenitor cells.* Stem Cells Dev, 2006. **15**(2): p. 245-53. DOI: 10.1089/scd.2006.15.245

[570] Fuchs, J.R., et al., *Cartilage engineering from ovine umbilical cord blood mesenchymal progenitor cells.* Stem Cells, 2005. **23**(7): p. 958-64. DOI: 10.1634/stemcells.2004-0310

[571] Philp, D., et al., *Complex extracellular matrices promote tissue-specific stem cell differentiation.* Stem Cells, 2005. **23**(2): p. 288-96. DOI: 10.1634/stemcells.2002-0109

[572] Pei, M., et al., *Bioreactors mediate the effectiveness of tissue engineering scaffolds.* Faseb J, 2002. **16**(12): p. 1691-4.

[573] Hu, J.C. and K.A. Athanasiou, *Low-density cultures of bovine chondrocytes: effects of scaffold material and culture system.* Biomaterials, 2005. **26**(14): p. 2001-12. DOI: 10.1016/j.biomaterials.2004.06.038

[574] Emin, N., et al., *Engineering of rat articular cartilage on porous sponges: effects of tgf-beta 1 and microgravity bioreactor culture.* Artif Cells Blood Substit Immobil Biotechnol, 2008. **36**(2): p. 123-37.

[575] Nettles, D.L., S.H. Elder, and J.A. Gilbert, *Potential use of chitosan as a cell scaffold material for cartilage tissue engineering.* Tissue Eng, 2002. **8**(6): p. 1009-16. DOI: 10.1089/107632702320934100

[576] Griffon, D.J., et al., *Chitosan scaffolds: interconnective pore size and cartilage engineering.* Acta Biomater, 2006. **2**(3): p. 313-20. DOI: 10.1016/j.actbio.2005.12.007

[577] Kasahara, Y., et al., *Development of mature cartilage constructs using novel three-dimensional porous scaffolds for enhanced repair of osteochondral defects.* J Biomed Mater Res A, 2008. **86**(1): p. 127-36.

[578] Chen, H.C., et al., *A novel rotating-shaft bioreactor for two-phase cultivation of tissue-engineered cartilage.* Biotechnol Prog, 2004. **20**(6): p. 1802-9. DOI: 10.1021/bp049740s

[579] Lagana, K., et al., *A new bioreactor for the controlled application of complex mechanical stimuli for cartilage tissue engineering.* Proc Inst Mech Eng [H], 2008. **222**(5): p. 705-15.

[580] Moretti, M., et al., *An integrated experimental-computational approach for the study of engineered cartilage constructs subjected to combined regimens of hydrostatic pressure and interstitial perfusion.* Biomed Mater Eng, 2008. **18**(4-5): p. 273-8.

[581] Carver, S.E. and C.A. Heath, *Influence of intermittent pressure, fluid flow, and mixing on the regenerative properties of articular chondrocytes.* Biotechnol Bioeng, 1999. **65**(3): p. 274-81. DOI: 10.1002/(SICI)1097-0290(19991105)65:3<274::AID-BIT4>3.0.CO;2-E

[582] French, M.M., et al., *Chondrogenic differentiation of adult dermal fibroblasts.* Ann Biomed Eng, 2004. **32**(1): p. 50-6. DOI: 10.1023/B:ABME.0000007790.65773.e0

[583] Schultz, S.S., S. Abraham, and P.A. Lucas, *Stem cells isolated from adult rat muscle differentiate across all three dermal lineages.* Wound Repair Regen, 2006. **14**(2): p. 224-31. DOI: 10.1111/j.1743-6109.2006.00114.x

[584] Heng, B.C., T. Cao, and E.H. Lee, *Directing stem cell differentiation into the chondrogenic lineage in vitro.* Stem Cells, 2004. **22**(7): p. 1152-67. DOI: 10.1634/stemcells.2004-0062

[585] Glowacki, J., E. Trepman, and J. Folkman, *Cell shape and phenotypic expression in chondrocytes.* Proceedings of the Society for Experimental Biology and Medicine. Society for Experimental Biology and Medicine (New York, N. Y.), 1983. **172(1)**: p. 93-8.

[586] von der Mark, K., et al., *Relationship between cell shape and type of collagen synthesized as chondrocytes lose their cartilage phenotype in culture.* Nature, 1977. **267**: p. 531-2.

[587] Betre, H., et al., *Chondrocytic differentiation of human adipose-derived adult stem cells in elastin-like polypeptide.* Biomaterials., 2006. **27**(1): p. 91-9. DOI: 10.1016/j.biomaterials.2005.05.071

[588] Nakayama, N., et al., *Macroscopic cartilage formation with embryonic stem-cell-derived mesodermal progenitor cells.* J Cell Sci, 2003. **116**(Pt 10): p. 2015-28. DOI: 10.1242/jcs.00417

[589] Sui, Y., T. Clarke, and J.S. Khillan, *Limb bud progenitor cells induce differentiation of pluripotent embryonic stem cells into chondrogenic lineage.* Differentiation, 2003. **71**(9-10): p. 578-85. DOI: 10.1111/j.1432-0436.2003.07109001.x

[590] zur Nieden, N.I., et al., *Induction of chondro-, osteo- and adipogenesis in embryonic stem cells by bone morphogenetic protein-2: effect of cofactors on differentiating lineages.* BMC Dev Biol, 2005. **5**(1): p. 1.

[591] Kramer, J., et al., *Embryonic stem cell-derived chondrogenic differentiation in vitro: activation by BMP-2 and BMP-4.* Mech Dev, 2000. **92**(2): p. 193-205. DOI: 10.1016/S0925-4773(99)00339-1

[592] Tanaka, H., et al., *Chondrogenic differentiation of murine embryonic stem cells: effects of culture conditions and dexamethasone.* J Cell Biochem, 2004. **93**(3): p. 454-62. DOI: 10.1002/jcb.20171

[593] Hegert, C., et al., *Differentiation plasticity of chondrocytes derived from mouse embryonic stem cells.* J Cell Sci, 2002. **115**(Pt 23): p. 4617-28. DOI: 10.1242/jcs.00171

[594] Kim, M.S., et al., *Musculoskeletal differentiation of cells derived from human embryonic germ cells.* Stem Cells, 2005. **23**(1): p. 113-23. DOI: 10.1634/stemcells.2004-0110

[595] Wulf, G.G., et al., *Mesengenic progenitor cells derived from human placenta.* Tissue Engineering, 2004. **10**(7-8): p. 1136-47.

[596] Thomson, J.A., et al., *Embryonic stem cell lines derived from human blastocysts.* Science., 1998. **282**(5391): p. 1145-7.

[597] Plaia, T.W., et al., *Characterization of a new NIH registered variant human embryonic stem cell line BG01V: A tool for human embryonic stem cell research.* Stem Cells, 2005.

[598] Zeng, X., et al., *BG01V: a variant human embryonic stem cell line which exhibits rapid growth after passaging and reliable dopaminergic differentiation.* Restor Neurol Neurosci, 2004. **22**(6): p. 421-8.

[599] Barberi, T., et al., *Derivation of multipotent mesenchymal precursors from human embryonic stem cells.* PLoS Med, 2005. **2**(6): p. e161. DOI: 10.1371/journal.pmed.0020161

[600] Amit, M., et al., *Feeder layer- and serum-free culture of human embryonic stem cells.* Biol Reprod, 2004. **70**(3): p. 837-45.

[601] Li, Y., et al., *Expansion of human embryonic stem cells in defined serum-free medium devoid of animal-derived products*. Biotechnol Bioeng, 2005. **91**(6): p. 688-98. DOI: 10.1002/bit.20536

[602] Amit, M., et al., *Human feeder layers for human embryonic stem cells*. Biol Reprod, 2003. **68**(6): p. 2150-6.

[603] Klimanskaya, I., et al., *Human embryonic stem cells derived without feeder cells*. Lancet, 2005. **365**(9471): p. 1631-41.

[604] Alhadlaq, A., et al., *Adult stem cell driven genesis of human-shaped articular condyle*. Ann Biomed Eng, 2004. **32**(7): p. 911-23. DOI: 10.1023/B:ABME.0000032454.53116.ee

[605] Alhadlaq, A. and J.J. Mao, *Tissue-engineered osteochondral constructs in the shape of an articular condyle*. J Bone Joint Surg Am, 2005. **87**(5): p. 936-44. DOI: 10.2106/JBJS.D.02104

[606] Mao, J.J., *Stem-cell-driven regeneration of synovial joints*. Biol Cell, 2005. **97**(5): p. 289-301.

[607] Bosnakovski, D., et al., *Chondrogenic differentiation of bovine bone marrow mesenchymal stem cells in pellet cultural system*. Exp Hematol, 2004. **32**(5): p. 502-9. DOI: 10.1016/j.exphem.2004.02.009

[608] Kramer, J., C. Hegert, and J. Rohwedel, *In vitro differentiation of mouse ES cells: bone and cartilage*. Methods In Enzymology, 2003. **365**: p. 251-268. DOI: 10.1016/S0076-6879(03)65018-4

[609] Noel, D., et al., *Short term BMP-2 expression is sufficient for in vivo osteo-chondral differentiation of mesenchymal stem cells*. Stem Cells, 2004. **22**: p. 74-85. DOI: 10.1634/stemcells.22-1-74

[610] Gelse, K., et al., *Articular cartilage repair by gene therapy using growth factor producing mesenchymal cells*. Arthritis and rheumatism., 2003. **48**: p. 430-441. DOI: 10.1002/art.10759

[611] van Beuningen, H.M., et al., *Differential effects of local application of BMP-2 or TGF-B1 on both articular cartilage composition and osteophyte formation*. Osteoarthritis and cartilage / OARS, Osteoarthritis Research Society., 1998. **6**: p. 306-317.

[612] Grande, D.A., et al., *Stem cells as platforms for delivery of genes to enhance cartilage repair*. J Bone Joint Surg Am, 2003. **85**(Suppl 2): p. 111-116.

[613] Tsuchiya, H., et al., *Chondrogenesis enhanced by overexpression of sox9 gene in mouse bone marrow-derived mesenchymal stem cells*. Biochem Biophys Res Commun, 2003. **301**: p. 338-343. DOI: 10.1016/S0006-291X(02)03026-7

[614] Jorgensen, C., J. Gordeladze, and D. Noel, *Tissue engineering through autologous mesenchymal stem cells*. Current Opinion in Biotechnology, 2004. **15**: p. 406-410. DOI: 10.1016/j.copbio.2004.08.003

[615] Erickson, G.R., et al., *Chondrogenic potential of adipose tissue-derived stromal cells in vitro and in vivo.* Biochem Biophys Res Commun, 2002. **290**(2): p. 763-9. DOI: 10.1006/bbrc.2001.6270

[616] Mehlhorn, A.T., et al., *Differential effects of BMP-2 and TGF-beta1 on chondrogenic differentiation of adipose derived stem cells.* Cell Prolif, 2007. **40**(6): p. 809-23. DOI: 10.1111/j.1365-2184.2007.00473.x

[617] Toma, J.G., et al., *Isolation and characterization of multipotent skin-derived precursors from human skin.* Stem Cells, 2005. **23**(6): p. 727-37. DOI: 10.1634/stemcells.2004-0134

[618] Toma, J.G., et al., *Isolation of multipotent adult stem cells from the dermis of mammalian skin.* Nat Cell Biol, 2001. **3**(9): p. 778-84. DOI: 10.1038/ncb0901-778

[619] Mizuno, S. and J. Glowacki, *Low oxygen tension enhances chondroinduction by demineralized bone matrix in human dermal fibroblasts in vitro.* Cells Tissues Organs, 2005. **180**(3): p. 151-8. DOI: 10.1159/000088243

[620] Glowacki, J., et al., *In vitro engineering of cartilage: effects of serum substitutes, TGF-beta, and IL-1alpha.* Orthod Craniofac Res, 2005. **8**(3): p. 200-8. DOI: 10.1111/j.1601-6343.2005.00333.x

[621] Hwang, N.S., et al., *The effects of three dimensional culture and growth factors on the chondrogenic differentiation of murine embryonic stem cells.* Stem Cells, 2005.

[622] Wong, M. and R.S. Tuan, *Nuserum, a synthetic serum replacement, supports chondrogenesis of embryonic chick limb bud mesenchymal cells in micromass culture.* In Vitro Cell Dev Biol Anim, 1993. **29A**: p. 917-22. DOI: 10.1007/BF02634229

[623] Lennon, D.P., et al., *A chemically defined medium supports in vitro proliferation and maintains the osteochondral potential of rat marrow-derived mesenchymal stem cells.* Experimental cell research., 1995. **219**: p. 211-22. DOI: 10.1006/excr.1995.1221

[624] Schmidt, C.C., et al., *Effects of growth factors on the proliferation of fibroblasts from the medial collateral and anterior cruciate ligaments.* J Orthop Res, 1995. **13**: p. 184-90. DOI: 10.1002/jor.1100130206

[625] Takahashi, K., et al., *Induction of pluripotent stem cells from adult human fibroblasts by defined factors.* Cell, 2007. **131**(5): p. 861-72.

[626] Yu, J., et al., *Induced pluripotent stem cell lines derived from human somatic cells.* Science, 2007. **318**(5858): p. 1917-20.

[627] Osafune, K., et al., *Marked differences in differentiation propensity among human embryonic stem cell lines.* Nat Biotechnol, 2008. **26**(3): p. 313-5. DOI: 10.1038/nbt1383

[628] Lowry, W.E., et al., *Generation of human induced pluripotent stem cells from dermal fibroblasts.* Proc Natl Acad Sci U S A, 2008. **105**(8): p. 2883-8. DOI: 10.1073/pnas.0711983105

[629] Stadtfeld, M., et al., *Induced Pluripotent Stem Cells Generated Without Viral Integration.* Science, 2008. DOI: 10.1126/science.1162494

[630] Okita, K., et al., *Generation of Mouse Induced Pluripotent Stem Cells Without Viral Vectors.* Science, 2008. DOI: 10.1126/science.1164270

[631] Nieminen, M.T., et al., *Spatial assessment of articular cartilage proteoglycans with Gd-DTPA-enhanced T1 imaging.* Magn Reson Med, 2002. **48**(4): p. 640-8. DOI: 10.1002/mrm.10273

[632] Juras, V., et al., *Kinematic biomechanical assessment of human articular cartilage transplants in the knee using 3-T MRI: an in vivo reproducibility study.* Eur Radiol, 2009. **19**(5): p. 1246-52. DOI: 10.1007/s00330-008-1242-0

[633] Juras, V., et al., *In vitro determination of biomechanical properties of human articular cartilage in osteoarthritis using multi-parametric MRI.* J Magn Reson, 2009. **197**(1): p. 40-7. DOI: 10.1016/j.jmr.2008.11.019

[634] Zalewski, T., et al., *Scaffold-aided repair of articular cartilage studied by MRI.* MAGMA, 2008. **21**(3): p. 177-85. DOI: 10.1007/s10334-008-0108-4

[635] Qin, L., et al., *Ultrasound detection of trypsin-treated articular cartilage: its association with cartilaginous proteoglycans assessed by histological and biochemical methods.* J Bone Miner Metab, 2002. **20**(5): p. 281-7. DOI: 10.1007/s007740200040

[636] Huang, Y.P. and Y.P. Zheng, *Intravascular Ultrasound (IVUS): A Potential Arthroscopic Tool for Quantitative Assessment of Articular Cartilage.* Open Biomed Eng J, 2009. **3**: p. 13-20. DOI: 10.2174/1874120700903010013

[637] Kuroki, H., et al., *Ultrasound has the potential to detect degeneration of articular cartilage clinically, even if the information is obtained from an indirect measurement of intrinsic physical characteristics.* Arthritis Res Ther, 2009. **11**(3): p. 408. DOI: 10.1186/ar2727

[638] Kisiday, J., A. Kerin, and A.J. Grodzinsky, *Mechanical Testing of Cell-Material Constructs: A Review,* in *Biopolymer Methods in Tissue Engineering,* A.P. Hollander and P.V. Hatton, Editors. 2003, Humana Press: Totowa, New Jersey. p. 239-54.

[639] Niederauer, G.G., et al. *Applications and advantages of a hand-held indentation device.* in *2002 International Cartilage Repair Society Symposium "Biophysical Diagnosis of Cartilage Degeneration and Repair".* June 15-18, 2002. Toronto, Canada.

[640] Niederauer, G.G., et al. *Sensitivity of a hand-held indentation probe for measuring the stiffness of articular cartilage.* in *The Second Joint Meeting of the IEEE Engineering in Medicine and Biology Society and the Biomedical Engineering Society.* 2002. Houston, Texas, USA.

[641] Toyras, J., et al., *Estimation of the Young's modulus of articular cartilage using an arthroscopic indentation instrument and ultrasonic measurement of tissue thickness.* J Biomech, 2001. **34**(2): p. 251-6. DOI: 10.1016/S0021-9290(00)00189-5

[642] Mak, A.F., W.M. Lai, and V.C. Mow, *Biphasic indentation of articular cartilage–I. Theoretical analysis.* J Biomech, 1987. **20**(7): p. 703-14. DOI: 10.1016/0021-9290(87)90036-4

[643] DiSilvestro, M.R., et al., *Biphasic poroviscoelastic simulation of the unconfined compression of articular cartilage: I–Simultaneous prediction of reaction force and lateral displacement.* J Biomech Eng, 2001. **123**(2): p. 191-7. DOI: 10.1115/1.1351890

[644] Guilak, F., et al., *Mechanical and biochemical changes in the superficial zone of articular cartilage in canine experimental osteoarthritis.* J Orthop Res, 1994. **12**(4): p. 474-84. DOI: 10.1002/jor.1100120404

[645] Stading, M. and R. Langer, *Mechanical shear properties of cell-polymer cartilage constructs.* Tissue Eng, 1999. **5**(3): p. 241-50. DOI: 10.1089/ten.1999.5.241

[646] Anderson, D.R., et al., *Viscoelastic shear properties of the equine medial meniscus.* J Orthop Res, 1991. **9**(4): p. 550-8. DOI: 10.1002/jor.1100090411

[647] Mow, V.C., G.A. Ateshian, and R.L. Spilker, *Biomechanics of diarthrodial joints: a review of twenty years of progress.* J Biomech Eng, 1993. **115**(4B): p. 460-7. DOI: 10.1115/1.2895525

[648] Katta, J., et al., *Biotribology of articular cartilage–a review of the recent advances.* Med Eng Phys, 2008. **30**(10): p. 1349-63. DOI: 10.1016/j.medengphy.2008.09.004

[649] Walker, P.S., et al., *"Boosted lubrication" in synovial joints by fluid entrapment and enrichment.* Ann Rheum Dis, 1968. **27**(6): p. 512-20. DOI: 10.1136/ard.27.6.512

[650] Linn, F.C., *Lubrication of animal joints. II. The mechanism.* J Biomech, 1968. **1**(3): p. 193-205. DOI: 10.1016/0021-9290(68)90004-3

[651] Jay, G.D., *Characterization of a bovine synovial fluid lubricating factor. I. Chemical, surface activity and lubricating properties.* Connect Tissue Res, 1992. **28**(1-2): p. 71-88.

[652] Park, S., K.D. Costa, and G.A. Ateshian, *Microscale frictional response of bovine articular cartilage from atomic force microscopy.* J Biomech, 2004. **37**(11): p. 1679-87. DOI: 10.1016/j.jbiomech.2004.02.017

[653] Ker, R.F., *The design of soft collagenous load-bearing tissues.* J Exp Biol, 1999. **202**(Pt 23): p. 3315-24.

[654] Simon, W.H., *Wear properties of articular cartilage in vitro.* J Biomech, 1971. **4**(5): p. 379-89. DOI: 10.1016/0021-9290(71)90058-3

[655] Frank, E.H. and A.J. Grodzinsky, *Cartilage electromechanics–II. A continuum model of cartilage electrokinetics and correlation with experiments.* J Biomech, 1987. **20**(6): p. 629-39. DOI: 10.1016/0021-9290(87)90283-1

[656] Lai, W.M., J.S. Hou, and V.C. Mow, *A triphasic theory for the swelling and deformation behaviors of articular cartilage.* J Biomech Eng, 1991. **113**(3): p. 245-58. DOI: 10.1115/1.2894880

[657] Jay, G.D., et al., *Lubricating ability of aspirated synovial fluid from emergency department patients with knee joint synovitis.* J Rheumatol, 2004. **31**(3): p. 557-64.

[658] Below, S., et al., *The split-line pattern of the distal femur: A consideration in the orientation of autologous cartilage grafts.* Arthroscopy, 2002. **18**(6): p. 613-7.

[659] van de Breevaart Bravenboer, J., et al., *Improved cartilage integration and interfacial strength after enzymatic treatment in a cartilage transplantation model.* Arthritis Res Ther, 2004. **6**(5): p. R469-76.

[660] Peretti, G.M., et al., *Cell-based bonding of articular cartilage: An extended study.* J Biomed Mater Res A, 2003. **64**(3): p. 517-24.

[661] Bos, P.K., et al., *Specific enzymatic treatment of bovine and human articular cartilage: implications for integrative cartilage repair.* Arthritis Rheum, 2002. **46**(4): p. 976-85. DOI: 10.1002/art.10208

[662] Hunziker, E.B., I.M. Driesang, and C. Saager, *Structural barrier principle for growth factor-based articular cartilage repair.* Clin Orthop Relat Res, 2001(391 Suppl): p. S182-9.

[663] Tew, S.R., et al., *The reactions of articular cartilage to experimental wounding: role of apoptosis.* Arthritis Rheum, 2000. **43**(1): p. 215-25. DOI: 10.1002/1529-0131(200001)43:1<215::AID-ANR26>3.0.CO;2-X

[664] Obradovic, B., et al., *Integration of engineered cartilage.* J Orthop Res, 2001. **19**(6): p. 1089-97. DOI: 10.1016/S0736-0266(01)00030-4

[665] Reindel, E.S., et al., *Integrative repair of articular cartilage in vitro: adhesive strength of the interface region.* J Orthop Res, 1995. **13**(5): p. 751-60. DOI: 10.1002/jor.1100130515

[666] Ahsan, T. and R.L. Sah, *Biomechanics of integrative cartilage repair.* Osteoarthritis Cartilage, 1999. **7**(1): p. 29-40. DOI: 10.1053/joca.1998.0160

[667] DiMicco, M.A., et al., *Integrative articular cartilage repair: dependence on developmental stage and collagen metabolism.* Osteoarthritis Cartilage, 2002. **10**(3): p. 218-25. DOI: 10.1053/joca.2001.0502

[668] Ahsan, T., et al., *Integrative cartilage repair: inhibition by beta-aminopropionitrile.* J Orthop Res, 1999. **17**(6): p. 850-7. DOI: 10.1002/jor.1100170610

[669] Peretti, G.M., et al., *Biomechanical analysis of a chondrocyte-based repair model of articular cartilage.* Tissue Eng, 1999. **5**(4): p. 317-26. DOI: 10.1089/ten.1999.5.317

[670] Wakitani, S., et al., *Mesenchymal cell-based repair of large, full-thickness defects of articular cartilage.* J Bone Joint Surg Am, 1994. **76**(4): p. 579-92.

[671] Yang, W.D., et al., *A study of injectable tissue-engineered autologous cartilage.* Chin J Dent Res, 2000. **3**(4): p. 10-5.

[672] Hunziker, E.B. and L.C. Rosenberg, *Repair of partial-thickness defects in articular cartilage: cell recruitment from the synovial membrane.* J Bone Joint Surg Am, 1996. **78**(5): p. 721-33.

[673] Hunziker, E.B. and E. Kapfinger, *Removal of proteoglycans from the surface of defects in articular cartilage transiently enhances coverage by repair cells.* J Bone Joint Surg Am, 1998. **80-B**: p. 144-150. DOI: 10.1302/0301-620X.80B1.7531

[674] Englert, C., et al., *Inhibition of integrative cartilage repair by proteoglycan 4 in synovial fluid.* Arthritis Rheum, 2005. **52**(4): p. 1091-9. DOI: 10.1002/art.20986

[675] Arem, A., *Collagen modifications.* Clin Plast Surg, 1985. **12**(2): p. 209-20.

[676] Wright, G.C., Jr., et al., *Stimulation of matrix formation in rabbit chondrocyte cultures by ascorbate. 1. Effect of ascorbate analogs and beta-aminopropionitrile.* J Orthop Res, 1988. **6**(3): p. 397-407. DOI: 10.1002/jor.1100060311

[677] Gerstenfeld, L.C., et al., *Post-translational control of collagen fibrillogenesis in mineralizing cultures of chick osteoblasts.* J Bone Miner Res, 1993. **8**(9): p. 1031-43.

[678] Samuels, J., S. Krasnokutsky, and S.B. Abramson, *Osteoarthritis: a tale of three tissues.* Bull NYU Hosp Jt Dis, 2008. **66**(3): p. 244-50.

[679] Jenkinson, C.M., et al., *Effects of dietary intervention and quadriceps strengthening exercises on pain and function in overweight people with knee pain: randomised controlled trial.* BMJ, 2009. **339**: p. b3170. DOI: 10.1136/bmj.b3170

[680] Teichtahl, A.J., et al., *The longitudinal relationship between body composition and patella cartilage in healthy adults.* Obesity (Silver Spring), 2008. **16**(2): p. 421-7. DOI: 10.1038/oby.2007.37

[681] Christensen, R., A. Astrup, and H. Bliddal, *Weight loss: the treatment of choice for knee osteoarthritis? A randomized trial.* Osteoarthritis Cartilage, 2005. **13**(1): p. 20-7. DOI: 10.1016/j.joca.2004.10.008

[682] Walker, W.R. and D.M. Keats, *An investigation of the therapeutic value of the 'copper bracelet'-dermal assimilation of copper in arthritic/rheumatoid conditions.* Agents Actions, 1976. **6**(4): p. 454-9.

[683] Richmond, S.J., *Magnet therapy for the relief of pain and inflammation in rheumatoid arthritis (CAMBRA): A randomised placebo-controlled crossover trial.* Trials, 2008. **9**: p. 53. DOI: 10.1186/1745-6215-9-53

[684] Zhang, W., et al., *OARSI recommendations for the management of hip and knee osteoarthritis, part I: critical appraisal of existing treatment guidelines and systematic review of current research evidence.* Osteoarthritis Cartilage, 2007. **15**(9): p. 981-1000. DOI: 10.1016/j.joca.2007.06.014

[685] Jordan, K.M., et al., *EULAR Recommendations 2003: an evidence based approach to the management of knee osteoarthritis: Report of a Task Force of the Standing Committee for International Clinical Studies Including Therapeutic Trials (ESCISIT).* Ann Rheum Dis, 2003. **62**(12): p. 1145-55. DOI: 10.1136/ard.2003.011742

[686] *The NIH Glucosamine/Chondroitin Arthritis Intervention Trial (GAIT).* J Pain Palliat Care Pharmacother, 2008. **22**(1): p. 39-43.

[687] Lee, Y.H., et al., *Effect of glucosamine or chondroitin sulfate on the osteoarthritis progression: a meta-analysis.* Rheumatol Int, 2009. DOI: 10.1007/s00296-009-0969-5

[688] Johnson, L.L., *Arthroscopic abrasion arthroplasty historical and pathologic perspective: present status.* Arthroscopy, 1986. **2**(1): p. 54-69.

[689] Federico, D.J. and B. Reider, *Results of isolated patellar debridement for patellofemoral pain in patients with normal patellar alignment.* Am J Sports Med, 1997. **25**(5): p. 663-9. DOI: 10.1177/036354659702500513

[690] Levy, A.S., et al., *Chondral delamination of the knee in soccer players.* Am J Sports Med, 1996. **24**(5): p. 634-9. DOI: 10.1177/036354659602400512

[691] Chiu, F.Y. and C.M. Chen, *Surgical debridement and parenteral antibiotics in infected revision total knee arthroplasty.* Clin Orthop Relat Res, 2007. **461**: p. 130-5. DOI: 10.1097/BLO.0b013e318063e7f3

[692] Jackson, R.W., J.E. Gilbert, and P.F. Sharkey, *Arthroscopic debridement versus arthroplasty in the osteoarthritic knee.* J Arthroplasty, 1997. **12**(4): p. 465-9; discussion 469-70.

[693] Vingerhoeds, B., I. Degreef, and L. De Smet, *Debridement arthroplasty for osteoarthritis of the elbow (Outerbridge-Kashiwagi procedure).* Acta Orthop Belg, 2004. **70**(4): p. 306-10.

[694] Wada, T., et al., *Debridement arthroplasty for primary osteoarthritis of the elbow. Surgical technique.* J Bone Joint Surg Am, 2005. **87**(Pt 1): p. 95-105. DOI: 10.2106/JBJS.D.02684

[695] Gould, N. and A.B. Flick, *Post-fracture, late debridement resection arthroplasty of the ankle.* Foot Ankle, 1985. **6**(2): p. 70-82.

[696] Cuff, D.J., et al., *The treatment of deep shoulder infection and glenohumeral instability with debridement, reverse shoulder arthroplasty and postoperative antibiotics.* J Bone Joint Surg Br, 2008. **90**(3): p. 336-42.

[697] Mithoefer, K., et al., *Chondral resurfacing of articular cartilage defects in the knee with the microfracture technique. Surgical technique.* J Bone Joint Surg Am, 2006. **88**(Pt 2): p. 294-304. DOI: 10.2106/JBJS.F.00292

[698] Bae, D.K., K.H. Yoon, and S.J. Song, *Cartilage healing after microfracture in osteoarthritic knees.* Arthroscopy, 2006. **22**(4): p. 367-74.

[699] Siebold, R., S. Lichtenberg, and P. Habermeyer, *Combination of microfracture and periostal-flap for the treatment of focal full thickness articular cartilage lesions of the shoulder: a prospective study.* Knee Surg Sports Traumatol Arthrosc, 2003. **11**(3): p. 183-9.

[700] Becher, C. and H. Thermann, *Results of microfracture in the treatment of articular cartilage defects of the talus.* Foot Ankle Int, 2005. **26**(8): p. 583-9.

[701] Hoemann, C.D., et al., *Chitosan-glycerol phosphate/blood implants improve hyaline cartilage repair in ovine microfracture defects.* J Bone Joint Surg Am, 2005. **87**(12): p. 2671-86. DOI: 10.2106/JBJS.D.02536

[702] Breinan, H.A., et al., *Healing of canine articular cartilage defects treated with microfracture, a type-II collagen matrix, or cultured autologous chondrocytes.* J Orthop Res, 2000. **18**(5): p. 781-9. DOI: 10.1002/jor.1100180516

[703] Dorotka, R., et al., *Repair of articular cartilage defects treated by microfracture and a three-dimensional collagen matrix.* Biomaterials, 2005. **26**(17): p. 3617-29. DOI: 10.1016/j.biomaterials.2004.09.034

[704] Erggelet, C., et al., *Formation of cartilage repair tissue in articular cartilage defects pretreated with microfracture and covered with cell-free polymer-based implants.* J Orthop Res, 2009. DOI: 10.1002/jor.20879

[705] Zhang, X., et al., *The synergistic effects of microfracture, perforated decalcified cortical bone matrix and adenovirus-bone morphogenetic protein-4 in cartilage defect repair.* Biomaterials, 2008. **29**(35): p. 4616-29. DOI: 10.1016/j.biomaterials.2008.07.051

[706] Kuo, A.C., et al., *Microfracture and bone morphogenetic protein 7 (BMP-7) synergistically stimulate articular cartilage repair.* Osteoarthritis Cartilage, 2006. **14**(11): p. 1126-35. DOI: 10.1016/j.joca.2006.04.004

[707] Mithoefer, K., et al., *Clinical Efficacy of the Microfracture Technique for Articular Cartilage Repair in the Knee: An Evidence-Based Systematic Analysis.* Am J Sports Med, 2009. DOI: 10.1177/0363546508328414

[708] Kreuz, P.C., et al., *Is microfracture of chondral defects in the knee associated with different results in patients aged 40 years or younger?* Arthroscopy, 2006. **22**(11): p. 1180-6.

[709] Marcacci, M., et al., *Use of autologous grafts for reconstruction of osteochondral defects of the knee.* Orthopedics, 1999. **22**(6): p. 595-600.

[710] Hangody, L., et al., *Arthroscopic autogenous osteochondral mosaicplasty for the treatment of femoral condylar articular defects. A preliminary report.* Knee Surg Sports Traumatol Arthrosc, 1997. **5**(4): p. 262-7. DOI: 10.1007/s001670050061

[711] Hangody, L. and P. Fules, *Autologous osteochondral mosaicplasty for the treatment of full-thickness defects of weight-bearing joints: ten years of experimental and clinical experience.* J Bone Joint Surg Am, 2003. **85**(Suppl 2): p. 25-32.

[712] Dozin, B., et al., *Comparative evaluation of autologous chondrocyte implantation and mosaicplasty: a multicentered randomized clinical trial.* Clin J Sport Med, 2005. **15**(4): p. 220-6.

[713] Szerb, I., et al., *Mosaicplasty: long-term follow-up.* Bull Hosp Jt Dis, 2005. **63**(1-2): p. 54-62.

[714] Burks, R.T., et al., *The use of a single osteochondral autograft plug in the treatment of a large osteochondral lesion in the femoral condyle: an experimental study in sheep.* Am J Sports Med, 2006. **34**(2): p. 247-55. DOI: 10.1177/0363546505279914

[715] Bi, X., et al., *A novel method for determination of collagen orientation in cartilage by Fourier transform infrared imaging spectroscopy (FT-IRIS).* Osteoarthritis Cartilage, 2005. **13**(12): p. 1050-8. DOI: 10.1016/j.joca.2005.07.008

[716] Seigel, J.P. and J.A. Donlon, *Approval Letter - Carticel.* 1997. Rockville, MD: Public Health Service.

[717] Wood, J.J., et al., *Autologous cultured chondrocytes: adverse events reported to the United States Food and Drug Administration.* J Bone Joint Surg Am, 2006. **88**(3): p. 503-7. DOI: 10.2106/JBJS.E.00103

[718] Brittberg, M., et al., *Treatment of deep cartilage defects in the knee with autologous chondrocyte transplantation.* N Engl J Med, 1994. **331**(14): p. 889-95. DOI: 10.1056/NEJM199410063311401

[719] Horas, U., et al., *Autologous chondrocyte implantation and osteochondral cylinder transplantation in cartilage repair of the knee joint. A prospective, comparative trial.* J Bone Joint Surg Am, 2003. **85-A**(2): p. 185-92.

[720] Roberts, S., et al., *Immunohistochemical study of collagen types I and II and procollagen IIA in human cartilage repair tissue following autologous chondrocyte implantation.* Knee, 2009. DOI: 10.1016/j.knee.2009.02.004

[721] Kurkijarvi, J.E., et al., *Evaluation of cartilage repair in the distal femur after autologous chondrocyte transplantation using T2 relaxation time and dGEMRIC.* Osteoarthritis Cartilage, 2007. **15**(4): p. 372-8. DOI: 10.1016/j.joca.2006.10.001

[722] Vasara, A.I., et al., *Indentation stiffness of repair tissue after autologous chondrocyte transplantation.* Clin Orthop Relat Res, 2005(433): p. 233-42.

[723] Peterson, L., et al., *Autologous chondrocyte transplantation. Biomechanics and long-term durability.* Am J Sports Med, 2002. **30**(1): p. 2-12.

[724] Erggelet, C., et al., *[Matrix-augmented autologous chondrocyte implantation in the knee–arthroscopic technique].* Oper Orthop Traumatol, 2008. **20**(3): p. 199-207.

[725] Niemeyer, P., et al., *Autologous chondrocyte implantation for the treatment of retropatellar cartilage defects: clinical results referred to defect localisation.* Arch Orthop Trauma Surg, 2008. **128**(11): p. 1223-31. DOI: 10.1007/s00402-007-0413-9

[726] McNickle, A.G., et al., *Outcomes of Autologous Chondrocyte Implantation in a Diverse Patient Population.* Am J Sports Med, 2009. DOI: 10.1177/0363546509332258

[727] Kon, E., et al., *Arthroscopic second-generation autologous chondrocyte implantation compared with microfracture for chondral lesions of the knee: prospective nonrandomized study at 5 years.* Am J Sports Med, 2009. **37**(1): p. 33-41.

[728] Alfredson, H. and R. Lorentzon, *Superior results with continuous passive motion compared to active motion after periosteal transplantation. A retrospective study of human patella cartilage defect treatment.* Knee Surg Sports Traumatol Arthrosc, 1999. **7**(4): p. 232-8.

[729] Kim, H.K., et al., *Effects of continuous passive motion and immobilization on synovitis and cartilage degradation in antigen induced arthritis.* J Rheumatol, 1995. **22**(9): p. 1714-21.

[730] Athanasiou, K.A., et al., *Biomechanical modeling of repair articular cartilage: Effects of passive motion on osteochondral defects in monkey knee joints.* Tissue Eng, 1998. **4**(2): p. 185-195. DOI: 10.1089/ten.1998.4.185

[731] Lee, M.S., et al., *Protective effects of intermittent hydrostatic pressure on osteoarthritic chondrocytes activated by bacterial endotoxin in vitro.* J Orthop Res, 2003. **21**(1): p. 117-122. DOI: 10.1016/S0736-0266(02)00085-2

[732] Peterson, L., et al., *Two- to 9-year outcome after autologous chondrocyte transplantation of the knee.* Clin Orthop Relat Res, 2000(374): p. 212-34.

[733] Reinold, M.M., et al., *Current concepts in the rehabilitation following articular cartilage repair procedures in the knee.* J Orthop Sports Phys Ther, 2006. **36**(10): p. 774-94.

[734] Hambly, K., et al., *Autologous chondrocyte implantation postoperative care and rehabilitation: science and practice.* Am J Sports Med, 2006. **34**(6): p. 1020-38. DOI: 10.1177/0363546505281918

[735] Giannoni, P., et al., *Autologous chondrocyte implantation (ACI) for aged patients: development of the proper cell expansion conditions for possible therapeutic applications.* Osteoarthritis Cartilage, 2005. **13**(7): p. 589-600. DOI: 10.1016/j.joca.2005.02.015

[736] Mithofer, K., et al., *Articular cartilage repair in soccer players with autologous chondrocyte transplantation: functional outcome and return to competition.* Am J Sports Med, 2005. **33**(11): p. 1639-46. DOI: 10.1177/0363546505275647

[737] Fulkerson, J.P., et al., *Anteromedial tibial tubercle transfer without bone graft.* Am J Sports Med, 1990. **18**(5): p. 490-6; discussion 496-7. DOI: 10.1177/036354659001800508

[738] Beck, P.R., et al., *Trochlear contact pressures after anteromedialization of the tibial tubercle.* Am J Sports Med, 2005. **33**(11): p. 1710-5. DOI: 10.1177/0363546505278300

[739] Farr, J., *Autologous chondrocyte implantation improves patellofemoral cartilage treatment outcomes.* Clin Orthop Relat Res, 2007. **463**: p. 187-94.

[740] Rose, T., et al., *Autologous osteochondral mosaicplasty for treatment of a posttraumatic defect of the lateral tibial plateau: a case report with two-year follow-up.* J Orthop Trauma, 2005. **19**(3): p. 217-22.

[741] Maiotti, M., C. Massoni, and F. Allegra, *The Use of Poli-Hydrogel Cylindrical Implants to Treat Deep Chondral Defects of the Knee*, in *Arthroscopy Association of North America*. 2005: Vancouver, BC.

[742] McNickle, A.G., M.T. Provencher, and B.J. Cole, *Overview of existing cartilage repair technology.* Sports Med Arthrosc, 2008. **16**(4): p. 196-201. DOI: 10.1097/JSA.0b013e31818cdb82

[743] Falez, F. and F. Sciarretta, *Treatment of osteochondral symptomatic defects of the knee with Salu-Cartilage.* J Bone Joint Surg [Br], 2005. **87**(suppl II): p. 202.

[744] Elisseeff, J., et al., *Transdermal photopolymerization for minimally invasive implantation.* Proc Natl Acad Sci U S A, 1999. **96**(6): p. 3104-7. DOI: 10.1073/pnas.96.6.3104

[745] Schubert, T., et al., *Long-term effects of chondrospheres on cartilage lesions in an autologous chondrocyte implantation model as investigated in the SCID mouse model.* Int J Mol Med, 2009. **23**(4): p. 455-60.

[746] Schagemann, J.C., et al., *Cell-laden and cell-free biopolymer hydrogel for the treatment of osteochondral defects in a sheep model.* Tissue Eng Part A, 2009. **15**(1): p. 75-82. DOI: 10.1089/ten.tea.2008.0087

[747] Steinwachs, M., *New technique for cell-seeded collagen matrix-supported autologous chondrocyte transplantation.* Arthroscopy, 2009. **25**(2): p. 208-11.

[748] Mitani, G., et al., *The properties of bioengineered chondrocyte sheets for cartilage regeneration.* BMC Biotechnol, 2009. **9**(1): p. 17. DOI: 10.1186/1472-6750-9-17

[749] Niemeyer, P., et al., *Characteristic complications after autologous chondrocyte implantation for cartilage defects of the knee joint.* Am J Sports Med, 2008. **36**(11): p. 2091-9. DOI: 10.1177/0363546508322131

[750] Kreuz, P.C., et al., *Treatment of focal degenerative cartilage defects with polymer-based autologous chondrocyte grafts: four-year clinical results.* Arthritis Res Ther, 2009. **11**(2): p. R33. DOI: 10.1186/ar2638

[751] Anders, S., et al., *[Matrix-associated autologous chondrocyte transplantation (MACT). Minimally invasive technique in the knee].* Oper Orthop Traumatol, 2008. **20**(3): p. 208-19.

[752] Ferruzzi, A., et al., *Autologous chondrocyte implantation in the knee joint: open compared with arthroscopic technique. Comparison at a minimum follow-up of five years.* J Bone Joint Surg Am, 2008. **90**(suppl 4): p. 90-101. DOI: 10.2106/JBJS.H.00633

[753] Marcacci, M., et al., *Articular cartilage engineering with Hyalograft C: 3-year clinical results.* Clin Orthop Relat Res, 2005(435): p. 96-105.

[754] Marcacci, M., S. Zaffagnini, and E. Kon. *Outcomes and results of 2nd generation autologous chondrocyte implantation.* in *Annual Meeting of the American Orthopaedic Society for Sports Medicine.* 2005. Keystone, CO.

[755] Lee, J.I., et al., *A newly developed immunoisolated bioartificial pancreas with cell sheet engineering.* Cell Transplant, 2008. **17**(1-2): p. 51-9.

[756] Bolano, L. and J.A. Kopta, *The immunology of bone and cartilage transplantation.* Orthopedics, 1991. **14**(9): p. 987-96.

[757] Moskalewski, S., A. Hyc, and A. Osiecka-Iwan, *Immune response by host after allogeneic chondrocyte transplant to the cartilage.* Microsc Res Tech, 2002. **58**(1): p. 3-13. DOI: 10.1002/jemt.10110

[758] Bujia, J., et al., *Humoral immune response against minor collagens type IX and XI in patients with cartilage graft resorption after reconstructive surgery.* Ann Rheum Dis, 1994. **53**(4): p. 229-34. DOI: 10.1136/ard.53.4.229

[759] Dayer, E., et al., *Cartilage proteoglycan-induced arthritis in BALB/c mice. Antibodies that recognize human and mouse cartilage proteoglycan and can cause depletion of cartilage proteoglycan with little or no synovitis.* Arthritis Rheum, 1990. **33**(9): p. 1394-405. DOI: 10.1002/art.1780330912

[760] Glant, T.T., et al., *Critical roles of glycosaminoglycan side chains of cartilage proteoglycan (aggrecan) in antigen recognition and presentation.* J Immunol, 1998. **160**(8): p. 3812-9.

[761] Takagi, T. and H.E. Jasin, *Interactions between anticollagen antibodies and chondrocytes.* Arthritis Rheum, 1992. **35**(2): p. 224-30. DOI: 10.1002/art.1780350217

[762] Yablon, I.G., S. Cooperband, and D. Covall, *Matrix antigens in allografts. The humoral response.* Clin Orthop Relat Res, 1982(168): p. 243-51.

[763] Romaniuk, A., et al., *Rejection of cartilage formed by transplanted allogeneic chondrocytes: evaluation with monoclonal antibodies.* Transpl Immunol, 1995. **3**(3): p. 251-7. DOI: 10.1016/0966-3274(95)80032-8

[764] Lance, E.M., *Immunological reactivity towards chondrocytes in rat and man: relevance to autoimmune arthritis.* Immunol Lett, 1989. **21**(1): p. 63-73. DOI: 10.1016/0165-2478(89)90013-8

[765] Malejczyk, J., *Natural anti-chondrocyte cytotoxicity of normal human peripheral blood mononuclear cells.* Clin Immunol Immunopathol, 1989. **50**(Pt 1): p. 42-52. DOI: 10.1016/0090-1229(89)90220-1

[766] Malejczyk, J., et al., *Natural cell-mediated cytotoxic activity against isolated chondrocytes in the mouse.* Clin Exp Immunol, 1985. **59**(1): p. 110-6.

[767] Yamaga, K.M., et al., *Differentiation antigens of human articular chondrocytes and their tissue distribution as assessed by monoclonal antibodies.* J Autoimmun, 1994. **7**(2): p. 203-17. DOI: 10.1006/jaut.1994.1016

[768] Hale, D.A., *Basic transplantation immunology.* Surg Clin North Am, 2006. **86**(5): p. 1103-25, v. DOI: 10.1016/j.suc.2006.06.015

[769] Pietra, B.A., *Transplantation immunology 2003: simplified approach.* Pediatr Clin North Am, 2003. **50**(6): p. 1233-59. DOI: 10.1016/S0031-3955(03)00119-6

[770] Goldsby, R.A., et al., *Immunology.* 5th ed. 2003, New York: W. H. Freeman and Company.

[771] Trivedi, H.L., *Immunobiology of rejection and adaptation.* Transplant Proc, 2007. **39**(3): p. 647-52. DOI: 10.1016/j.transproceed.2007.01.047

[772] Langer, F. and A.E. Gross, *Immunogenicity of allograft articular cartilage.* J Bone Joint Surg Am, 1974. **56**(2): p. 297-304.

[773] Kaminski, M.J., G. Kaminska, and S. Moskalewski, *Species differences in the ability of isolated epiphyseal chondrocytes to hypertrophy after transplantation into the wall of the Syrian hamster cheek pouch.* Folia Biol (Krakow), 1980. **28**(1): p. 27-36.

[774] Ksiazek, T. and S. Moskalewski, *Studies on bone formation by cartilage reconstructed by isolated epiphyseal chondrocytes, transplanted syngeneically or across known histocompatibility barriers in mice.* Clin Orthop Relat Res, 1983(172): p. 233-42.

[775] Malejczyk, J., et al., *Effect of immunosuppression on rejection of cartilage formed by transplanted allogeneic rib chondrocytes in mice.* Clin Orthop Relat Res, 1991(269): p. 266-73.

[776] Malejczyk, J. and S. Moskalewski, *Effect of immunosuppression on survival and growth of cartilage produced by transplanted allogeneic epiphyseal chondrocytes.* Clin Orthop Relat Res, 1988(232): p. 292-303.

[777] Green, W.T., Jr., *Articular cartilage repair. Behavior of rabbit chondrocytes during tissue culture and subsequent allografting.* Clin Orthop Relat Res, 1977(124): p. 237-50.

[778] Glenn, R.E., Jr., et al., *Comparison of fresh osteochondral autografts and allografts: a canine model.* Am J Sports Med, 2006. **34**(7): p. 1084-93. DOI: 10.1177/0363546505284846

[779] Aston, J.E. and G. Bentley, *Repair of articular surfaces by allografts of articular and growth-plate cartilage.* J Bone Joint Surg Br, 1986. **68**(1): p. 29-35.

[780] Stevenson, S., et al., *The fate of articular cartilage after transplantation of fresh and cryopreserved tissue-antigen-matched and mismatched osteochondral allografts in dogs.* J Bone Joint Surg Am, 1989. **71**(9): p. 1297-307.

[781] Elves, M.W., *Humoral immune response to allografts of bone.* Int Arch Allergy Appl Immunol, 1974. **47**(5): p. 708-15.

[782] Friedlaender, G.E., *Immune responses to osteochondral allografts. Current knowledge and future directions.* Clin Orthop Relat Res, 1983(174): p. 58-68.

[783] Bakay, A., et al., *Osteochondral resurfacing of the knee joint with allograft. Clinical analysis of 33 cases.* Int Orthop, 1998. **22**(5): p. 277-81.

[784] Flynn, J.M., D.S. Springfield, and H.J. Mankin, *Osteoarticular allografts to treat distal femoral osteonecrosis.* Clin Orthop Relat Res, 1994(303): p. 38-43.

[785] Gross, A.E., N. Shasha, and P. Aubin, *Long-term followup of the use of fresh osteochondral allografts for posttraumatic knee defects.* Clin Orthop Relat Res, 2005(435): p. 79-87.

[786] Beaver, R.J., et al., *Fresh osteochondral allografts for post-traumatic defects in the knee. A survivorship analysis.* J Bone Joint Surg Br, 1992. **74**(1): p. 105-10.

[787] Chu, C.R., et al., *Articular cartilage transplantation. Clinical results in the knee.* Clin Orthop Relat Res, 1999(360): p. 159-68.

[788] Zukor, D.J., R.D. Oakeshott, and A.E. Gross, *Osteochondral allograft reconstruction of the knee. Part 2: Experience with successful and failed fresh osteochondral allografts.* Am J Knee Surg, 1989. **2**: p. 182.

[789] Masuoka, K., et al., *Tissue engineering of articular cartilage using an allograft of cultured chondrocytes in a membrane-sealed atelocollagen honeycomb-shaped scaffold (ACHMS scaffold).* J Biomed Mater Res B Appl Biomater, 2005. **75**(1): p. 177-84.

[790] Wakitani, S., et al., *Repair of rabbit articular surfaces with allograft chondrocytes embedded in collagen gel.* J Bone Joint Surg Br, 1989. **71**(1): p. 74-80.

[791] Wakitani, S., et al., *Repair of large full-thickness articular cartilage defects with allograft articular chondrocytes embedded in a collagen gel.* Tissue Eng, 1998. **4**(4): p. 429-44. DOI: 10.1089/ten.1998.4.429

[792] Rahfoth, B., et al., *Transplantation of allograft chondrocytes embedded in agarose gel into cartilage defects of rabbits.* Osteoarthritis Cartilage, 1998. **6**(1): p. 50-65. DOI: 10.1053/joca.1997.0092

[793] Schreiber, R.E., et al., *Repair of osteochondral defects with allogeneic tissue engineered cartilage implants.* Clin Orthop Relat Res, 1999(367 Suppl): p. S382-95.

[794] Butnariu-Ephrat, M., et al., *Resurfacing of goat articular cartilage by chondrocytes derived from bone marrow.* Clin Orthop Relat Res, 1996(330): p. 234-43.

[795] Osiecka-Iwan, A., et al., *Transplants of rat chondrocytes evoke strong humoral response against chondrocyte-associated antigen in rabbits.* Cell Transplant, 2003. **12**(4): p. 389-98.

[796] Ramallal, M., et al., *Xeno-implantation of pig chondrocytes into rabbit to treat localized articular cartilage defects: an animal model.* Wound Repair Regen, 2004. **12**(3): p. 337-45. DOI: 10.1111/j.1067-1927.2004.012309.x

[797] Costa, C., et al., *Transgenic pigs designed to express human CD59 and H-transferase to avoid humoral xenograft rejection.* Xenotransplantation, 2002. **9**(1): p. 45-57. DOI: 10.1034/j.1399-3089.2002.0o142.x

[798] Costa, C., J.L. Brokaw, and W.L. Fodor, *Characterization of cartilage from H-transferase transgenic pigs.* Transplant Proc, 2008. **40**(2): p. 554-6. DOI: 10.1016/j.transproceed.2007.12.025

[799] Li, W.J., et al., *Evaluation of articular cartilage repair using biodegradable nanofibrous scaffolds in a swine model: a pilot study.* J Tissue Eng Regen Med, 2009. **3**(1): p. 1-10. DOI: 10.1002/term.127

[800] Ren, G., et al., *Mesenchymal stem cell-mediated immunosuppression occurs via concerted action of chemokines and nitric oxide.* Cell Stem Cell, 2008. **2**(2): p. 141-50. DOI: 10.1016/j.stem.2007.11.014

[801] Gilbert, T.W., T.L. Sellaro, and S.F. Badylak, *Decellularization of tissues and organs.* Biomaterials, 2006. **27**(19): p. 3675-83.

[802] Chen, R.N., et al., *Process development of an acellular dermal matrix (ADM) for biomedical applications.* Biomaterials, 2004. **25**(13): p. 2679-86. DOI: 10.1016/j.biomaterials.2003.09.070

[803] Stapleton, T.W., et al., *Development and characterization of an acellular porcine medial meniscus for use in tissue engineering.* Tissue Eng Part A, 2008. **14**(4): p. 505-18. DOI: 10.1089/tea.2007.0233

[804] Lumpkins, S.B., N. Pierre, and P.S. McFetridge, *A mechanical evaluation of three decellularization methods in the design of a xenogeneic scaffold for tissue engineering the temporomandibular joint disc.* Acta Biomater, 2008. **4**(4): p. 808-16. DOI: 10.1016/j.actbio.2008.01.016

[805] Cartmell, J.S. and M.G. Dunn, *Effect of chemical treatments on tendon cellularity and mechanical properties.* J Biomed Mater Res, 2000. **49**(1): p. 134-40. DOI: 10.1002/(SICI)1097-4636(200001)49:1<134::AID-JBM17>3.0.CO;2-D

[806] Woods, T. and P.F. Gratzer, *Effectiveness of three extraction techniques in the development of a decellularized bone-anterior cruciate ligament-bone graft.* Biomaterials, 2005. **26**(35): p. 7339-49. DOI: 10.1016/j.biomaterials.2005.05.066

[807] Liao, J., E.M. Joyce, and M.S. Sacks, *Effects of decellularization on the mechanical and structural properties of the porcine aortic valve leaflet.* Biomaterials, 2008. **29**(8): p. 1065-74. DOI: 10.1016/j.biomaterials.2007.11.007

[808] Kasimir, M.T., et al., *Comparison of different decellularization procedures of porcine heart valves.* Int J Artif Organs, 2003. **26**(5): p. 421-7.

[809] Seebacher, G., et al., *Biomechanical properties of decellularized porcine pulmonary valve conduits.* Artif Organs, 2008. **32**(1): p. 28-35.

[810] Tudorache, I., et al., *Tissue engineering of heart valves: biomechanical and morphological properties of decellularized heart valves.* J Heart Valve Dis, 2007. **16**(5): p. 567-73; discussion 574.

[811] Grauss, R.W., et al., *Histological evaluation of decellularised porcine aortic valves: matrix changes due to different decellularisation methods.* Eur J Cardiothorac Surg, 2005. **27**(4): p. 566-71. DOI: 10.1016/j.ejcts.2004.12.052

[812] Meyer, S.R., et al., *Comparison of aortic valve allograft decellularization techniques in the rat.* J Biomed Mater Res A, 2006. **79**(2): p. 254-62.

[813] Meyer, S.R., et al., *Decellularization reduces the immune response to aortic valve allografts in the rat.* J Thorac Cardiovasc Surg, 2005. **130**(2): p. 469-76.

[814] Rosario, D.J., et al., *Decellularization and sterilization of porcine urinary bladder matrix for tissue engineering in the lower urinary tract.* Regen Med, 2008. **3**(2): p. 145-56. DOI: 10.2217/17460751.3.2.145

[815] Dahl, S.L., et al., *Decellularized native and engineered arterial scaffolds for transplantation.* Cell Transplant, 2003. **12**(6): p. 659-66.

[816] Hodde, J. and M. Hiles, *Virus safety of a porcine-derived medical device: evaluation of a viral inactivation method.* Biotechnol Bioeng, 2002. **79**(2): p. 211-6. DOI: 10.1002/bit.10281

[817] Hodde, J., et al., *Effects of sterilization on an extracellular matrix scaffold: part I. Composition and matrix architecture.* J Mater Sci Mater Med, 2007. **18**(4): p. 537-43. DOI: 10.1007/s10856-007-2301-9

[818] von Rechenberg, B., et al., *Changes in subchondral bone in cartilage resurfacing–an experimental study in sheep using different types of osteochondral grafts.* Osteoarthritis Cartilage, 2003. **11**(4): p. 265-77.

[819] Elder, B.D., S.V. Eleswarapu, and K.A. Athanasiou, *Evaluating Extraction Techniques for the Decellularization of Tissue Engineered Articular Cartilage Constructs.* Biomaterials, 2009: p. (In press). DOI: 10.1016/j.biomaterials.2009.03.050

[820] Elder, B.D., D.H. Kim, and K.A. Athanasiou, *Developing an Articular Cartilage Decellularization Process Towards Facet Joint Cartilage Replacement.* Neurosurgery, 2009 (Submitted).

[821] Auchincloss, H., Jr., *Xenogeneic transplantation. A review.* Transplantation, 1988. **46**(1): p. 1-20. DOI: 10.1097/00007890-198807000-00001

[822] Platt, J.L., et al., *Transplantation of discordant xenografts: a review of progress.* Immunol Today, 1990. **11**(12): p. 450-6; discussion 456-7.

[823] Jackson, D.W., et al., *Meniscal transplantation using fresh and cryopreserved allografts. An experimental study in goats.* Am J Sports Med, 1992. **20**(6): p. 644-56. DOI: 10.1177/036354659202000605

[824] Revell, C.M. and K.A. Athanasiou, *Success rates and immunologic responses of autologous, allogeneic, and xenogenic treatments to repair articular cartilage defects.* Tissue Eng Part B Rev, 2009. **15**(1): p. 1-15.

[825] Collins, B.H., et al., *Cardiac xenografts between primate species provide evidence for the importance of the alpha-galactosyl determinant in hyperacute rejection.* J Immunol, 1995. **154**(10): p. 5500-10.

[826] Galili, U., *Interaction of the natural anti-Gal antibody with alpha-galactosyl epitopes: a major obstacle for xenotransplantation in humans.* Immunol Today, 1993. **14**(10): p. 480-2. DOI: 10.1016/0167-5699(93)90261-I

[827] Good, A.H., et al., *Identification of carbohydrate structures that bind human antiporcine antibodies: implications for discordant xenografting in humans.* Transplant Proc, 1992. **24**(2): p. 559-62.

[828] Sandrin, M.S., et al., *Anti-pig IgM antibodies in human serum react predominantly with Gal(alpha 1-3)Gal epitopes.* Proc Natl Acad Sci U S A, 1993. **90**(23): p. 11391-5. DOI: 10.1073/pnas.90.23.11391

[829] Stone, K.R., et al., *Porcine and bovine cartilage transplants in cynomolgus monkey: I. A model for chronic xenograft rejection.* Transplantation, 1997. **63**(5): p. 640-5.

[830] Galili, U., et al., *Porcine and bovine cartilage transplants in cynomolgus monkey: II. Changes in anti-Gal response during chronic rejection.* Transplantation, 1997. **63**(5): p. 646-51.

[831] Stone, K.R., et al., *Porcine cartilage transplants in the cynomolgus monkey. III. Transplantation of alpha-galactosidase-treated porcine cartilage.* Transplantation, 1998. **65**(12): p. 1577-83.

[832] Tearle, R.G., et al., *The alpha-1,3-galactosyltransferase knockout mouse. Implications for xenotransplantation.* Transplantation, 1996. **61**(1): p. 13-9.

[833] Derham, C., et al., *Tissue engineering small-diameter vascular grafts: preparation of a biocompatible porcine ureteric scaffold.* Tissue Eng Part A, 2008. **14**(11): p. 1871-82. DOI: 10.1089/ten.tea.2007.0103

[834] *Compilation of Selected Acts within the Jurisdiction of the Committee on Energy and Commerce: Food, Drug, and Related Law, as Amended through December 31, 2004.* 109th Congress, 1st Session, 2005. Washington: U.S. GPO. GPO #: Y4.C73/8:107-D.

[835] Current good manufacturing practice in manufacturing, processing, packing, or holding of drugs. Title 21 Code of Federal Regulations, Pt. 210. Revised as of April 1, 2009.

[836] Current good manufacturing practice for finished pharmaceuticals. Title 21 Code of Federal Regulations, Pt. 211. Revised as of April 1, 2009.

[837] Quality system regulation. Title 21 Code of Federal Regulations, Pt. 820. Revised as of April 1, 2009.

[838] Investigational device exemptions. Title 21 Code of Federal Regulations, Pt. 812. Revised as of April 1, 2009.

[839] Trommelmans, L., J. Selling, and K. Dierickx, *A critical assessment of the directive on tissue engineering of the European union.* Tissue Eng, 2007. **13**(4): p. 667-72. DOI: 10.1089/ten.2006.0089

[840] Brevignon-Dodin, L. and F. Livesey, *Regulation of tissue-engineered products in the European Union: where are we heading?* Regen Med, 2006. **1**(5): p. 709-14. DOI: 10.2217/17460751.1.5.709

[841] McCulloch, P.C., et al., *Prospective evaluation of prolonged fresh osteochondral allograft transplantation of the femoral condyle: minimum 2-year follow-up.* Am J Sports Med, 2007. **35**(3): p. 411-20.

Authors' Biographies

K. A. ATHANASIOU

K. A. Athanasiou is a Distinguished Professor and the Chair of the Department of Biomedical Engineering at the University of California Davis. He holds a Ph.D. in mechanical engineering (bioengineering) from Columbia University.

E. M. DARLING

E. M. Darling is an Assistant Professor of Medical Science in the Department of Molecular Pharmacology, Physiology, & Biotechnology at Brown University. He holds a B.S. in engineering from Harvey Mudd College and a Ph.D. in bioengineering from Rice University.

J. C. HU

J. C. Hu is a Principal Development Engineer and Project Manager in the Department of Biomedical Engineering at the University of California Davis. He holds a B.S. in chemical engineering from The University of Texas at Austin and a Ph.D. in bioengineering from Rice University.